CAMBRIDGE LIBRARY COLLECTION

Books of enduring scholarly value

Darwin

Two hundred years after his birth and 150 years after the publication of 'On the Origin of Species', Charles Darwin and his theories are still the focus of worldwide attention. This series offers not only works by Darwin, but also the writings of his mentors in Cambridge and elsewhere, and a survey of the impassioned scientific, philosophical and theological debates sparked by his 'dangerous idea'.

History of Quadrupeds

Thomas Pennant (1726-98) was a keen geologist, naturalist and antiquary, who wrote a number of successful travel books about the British Isles as well as works on science. Linnaeus supported his election to the Royal Swedish Society of Sciences in 1757, and in 1767 he became a Fellow of the Royal Society. His work in zoology also earned him an honorary degree. His History of Quadrupeds (1793), aimed to promote natural history among a wider readership, originated in an informal index to John Ray's Synopsis of 1693. In his preface, Pennant acknowledges the monumental Histoire naturelle by the Comte de Buffon, as well as works by Klein (1751), Brisson (1756), and particularly the work of Linnaeus, though Pennant strongly disagreed with Linnaueus's classification of primates as including humans with apes. Pennant's two-volume book, beautifully illustrated with over 100 engravings, provides an accessible overview of the state of zoological classification at the end of the eighteenth century. Charles Darwin owned a copy and had it sent to him in South America during the Beagle voyage.

Cambridge University Press has long been a pioneer in the reissuing of out-of-print titles from its own backlist, producing digital reprints of books that are still sought after by scholars and students but could not be reprinted economically using traditional technology. The Cambridge Library Collection extends this activity to a wider range of books which are still of importance to researchers and professionals, either for the source material they contain, or as landmarks in the history of their academic discipline.

Drawing from the world-renowned collections in the Cambridge University Library, and guided by the advice of experts in each subject area, Cambridge University Press is using state-of-the-art scanning machines in its own Printing House to capture the content of each book selected for inclusion. The files are processed to give a consistently clear, crisp image, and the books finished to the high quality standard for which the Press is recognised around the world. The latest print-on-demand technology ensures that the books will remain available indefinitely, and that orders for single or multiple copies can quickly be supplied.

The Cambridge Library Collection will bring back to life books of enduring scholarly value (including out-of-copyright works originally issued by other publishers) across a wide range of disciplines in the humanities and social sciences and in science and technology.

History of Quadrupeds

VOLUME 1

THOMAS PENNANT

CAMBRIDGE UNIVERSITY PRESS

Cambridge, New York, Melbourne, Madrid, Cape Town, Singapore,
São Paolo, Delhi, Dubai, Tokyo

Published in the United States of America by Cambridge University Press, New York

www.cambridge.org
Information on this title: www.cambridge.org/9781108005166

This edition first published 1793
This digitally printed version 2009

ISBN 978-1-108-00516-6 Paperback

HISTORY

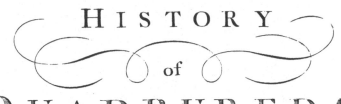

of

QUADRUPEDS,

The Third Edition.

Vol. I.

HEBDDUW HEB DDIM ADUW ADIGON.

PREFACE.

THE following work was originally intended for private amufement, and as an Index, for the more ready turning to any particular animal in the voluminous hiftory of quadrupeds by the late *Comte* DE BUFFON : But as it fwelled by degrees to a fize beyond my firft expectation, in the end I was determined to fling it into its prefent form, and to ufher it into the world.

THE prefent edition has prefumed to alter its title of SYNOPSIS to that of HISTORY ; not only on account of the vaft additions it has received, by favour of my friends, but likewife to prevent confufion among fuch who may think them worthy of the honor of quotation.

THE Synopfis of our illuftrious countryman, Mr. RAY, has been long out of print; and though, from his enlarged knowledge and great induftry one might well fuppofe his Work would for fome time difcourage all further attempts of the fame fort, yet a republication of that Synopfis would not have anfwered our prefent defign : For, living at a period when the ftudy of Natural Hiftory was but beginning to dawn in thefe Kingdoms, and when our contracted Commerce deprived him of many lights we now enjoy, he was obliged to content himfelf with giving defcriptions of the few Animals brought over here,

VOL. I.   a

and collecting the reft of his materials from other Writers. Yet
fo correct was his genius, that we view a fyftematic arrangement
arife even from the Chaos of *Aldrovandus* and *Gefner*. Under
his hand the indigefted matter of thefe able and copious Writers
affumes a new form, and the whole is made clear and perfpi-
cuous.

FROM this period every Writer on thefe fubjects propofed his
own method as an example; fome openly, but others more co-
vertly, aiming at the honor of originality, and attempting to
feek for fame in the path chalked out by Mr. RAY; but too
often without acknowleging the merit of the Guide.

MR. KLEIN, in 1751, made his appearance as a Syftematic
Writer on Quadrupeds, and in his firft order follows the general
arrangement of Mr. RAY; but the change he has made of feparating certain animals, which the laft had confolidated, are executed with great judgment. He feems lefs fortunate in his
fecond order; for, by a fervile regard to a method taken from
the number of toes, he has jumbled together moft oppofite animals; the *Camel* and the *Sloth*, the *Mole* and the *Bat*, the *Glutton*
and *Apes*; happy only in throwing back the *Walrus*, the *Seal*, and
the *Manati*, to the extremity of his fyftem: I fuppofe, as animals
nearly bordering on another clafs.

M. BRISSON, in 1756, favored the world with another fyftem,
arranging his animals by the number or defect of their teeth;
beginning with thofe that were toothlefs, fuch as the *Ant-eater*,
and ending with thofe that had the greateft number, fuch as the
Opoffum. By this method, laudable as it is in many refpects, it
muft happen unavoidably that fome quadrupeds, very diftant
from each other in their manners, are too clofely connected in

his Syftem; a defect which, however common, fhould be carefully avoided by every Naturalift.

In point of time, Linnæus ought to have the precedence; for he publifhed his firft Syftem in 1735. This was followed by feveral others, varying conftantly in the arrangement of the animal kingdom, even to the edition of 1766. It is, therefore, difficult to defend, and ftill more ungrateful to drop any reflections on a Naturalift, to whom we are fo greatly indebted. The variations in his different Syftems may have arifen from the new and continual difcoveries that are made in the animal kingdom; from his fincere intention of giving his Syftems additional improvements; and perhaps from a failing, (unknown indeed to many of his accufers) a diffidence in the abilities he had exerted in his prior performances. But it muft be allowed, that the Naturalift ran too great a hazard in imitating his *prefent* guife; for in another year he might put on a new form, and have left the complying Philofopher amazed at the metamorphofis.

But this is not my only reafon for rejecting the fyftem of this otherwife able Naturalift: There are faults in his arrangement of *Mammalia* *, that oblige me to feparate myfelf, in this one inftance, from his crowd of votaries; but that my feceffion may not appear the effect of whim or envy, it is to be hoped that the following objections will have their weight.

I reject his firft divifion, which he calls *Primates,* or Chiefs of the Creation; becaufe my vanity will not fuffer me to rank man-

---

* Or animals which have paps and fuckle their young; in which clafs are comprehended not only all the genuine quadrupeds, but even the Cetaceous tribe.

kind

kind with *Apes*, *Monkies*, *Maucaucos*, and *Bats*, the companions
LINNÆUS has allotted us even in his last System.

THE second order of *Bruta* I avoid for much the same reason:
The most intelligent of Quadrupeds, the half reasoning *Elephant*,
is made to associate with the most discordant and stupid of the
creation, with *Sloths*, *Ant-eaters*, and *Armadillos*, or with *Mana-
ties* and *Walruses*, inhabitants of another element.

THE third order of *Feræ* is not more admissible in all its articles; for it will be impossible to allow the *Mole*, the *Shrew*,
and the harmless *Hedge-hog*, to be the companions of *Lions*,
Wolves, and *Bears*: We may err in our arrangement.

Sed non ut placidis coeant immitia, non ut
Serpentes avibus geminentur, tigribus agni.

IN his arrangement of his fourth and fifth orders we quite
agree, except in the single article *Noctilio*, a species of Bat, which
happening to have only two cutting teeth in each jaw, is sepa-
rated from its companions, and placed with Squirrels, and others
of that class.

THE sixth order is made up of animals of the hoofed tribe;
but of genera so different in their nature, that notwithstanding
we admit them into the same division, we place them at such
distances from each other, with so many intervening links and
softening gradations, as will, it may be hoped, lessen the shock of
seeing the *Horse* and the *Hippopotame* in the same piece. To avoid
this as much as possible, we have flung the last into the back
ground, where it will appear more tolerable to the Critic, than
if they were left in a manner conjoined.

7

THE laſt order is that of Whales : which, it muſt be confeſſed, have, in many reſpects, the ſtructure of land animals; but their want of hair and feet, their fiſh-like form, and their conſtant reſidence in the water, are arguments for ſeparating them from this claſs, and forming them into another, independent of the reſt.

BUT while I thus freely offer my objections againſt embracing this Syſtem of Quadrupeds, let me not be ſuppoſed inſenſible of the other merits of this great and extraordinary perſon : His arrangement of fiſhes, of inſects, and of ſhells, are original and excellent ; he hath, in all his claſſes, given philoſophy a new language ; hath invented apt names, and taught the world a brevity, yet a fulneſs of deſcription, unknown to paſt ages : he hath with great induſtry brought numbers of ſynonyms of every animal into one point of view ; and hath given a conciſe account of the uſes and manners of each, as far as his obſervation extended, or the information of a numerous train of travelling diſciples could contribute : His Country may triumph in producing ſo vaſt a Genius, whoſe ſpirit invigorates ſcience in all that chilly region, and diffuſes it from thence to climates more favorable, which gratefully acknowledge the advantage of its influences.

LET us now turn our eyes to a Genius of another kind, to whom the Hiſtory of Quadrupeds owes very conſiderable lights : I mean the *Comte de Buffon*, who, in the moſt beautiful language, and in the moſt agreeable manner, hath given the ampleſt deſcriptions of the œconomy of the whole four-footed creation * : Such is his eloquence, that we forget the exuberant manner in which he treats each ſubject, and the reflections he often caſts on other

* For the anatomical part is the province of M. *D'Aubenton.*

Writers ;

Writers; the creation of his own gay fancy. Having in his own mind a comprehenfive view of every animal, he unfortunately feems to think it beneath him to fhackle his lively fpirit with fyftematic arrangement; fo that the Reader is forced to wander through numbers of volumes in fearch of any wifhed-for fubject. The mifunderftanding between thefe two able Naturalifts is moft injurious to fcience. The *French* Philofopher fcarcely mentions the *Swede*, but to treat him with contempt; *Linnæus*, in return, never deigns even to quote M. *de Buffon*, notwithftanding he muft know what ample lights he might have drawn from him.

I SHALL in a few words mention the plan that is followed in the prefent diftribution of quadrupeds, and at the fame time fhall clame but a fmall fhare of originality.

I COPY Mr. RAY, in his greater divifions of animals into hoofed, and digitated; but, after the manner of Mr. KLEIN, form feparate genera of the *Rhinoceros, Hippopotame, Tapiir,* and *Mufk.* The *Camel* being a ruminating animal, wanting the upper fore-teeth, and having the rudiments of hoofs, is placed in the firft order, after the *Mufk,* a hornlefs cloven-hoofed quadruped.

THE *Apes* are continued in the fame rank Mr. RAY has placed them, and are followed by the *Maucaucos.*

THE carnivorous animals deviate but little from his fyftem, and are arranged according to that of LINNÆUS, after omitting the *Seal, Mole, Shrew,* and *Hedge-hog.*

THE herbivorous or frugivorous quadrupeds keep here the fame ftation that our countryman affigned them; but this clafs comprehends befides, the *Shrew,* the *Mole,* and the *Hedge-hog.* The *Mole* is an exception to the character of this order, in refpect to the number of its cutting teeth; but its way of life, and its food, place it here more naturally than with the *Feræ*, as LIN-NÆUS has done. Thefe exceptions are to be met with even in the method * of that able Naturalift; nor can it be otherwife in all human fyftems; we are fo ignorant of many of the links of the chains of beings, that to expect perfection in the arrange-ment of them, would be the moft weak prefumption. We ought, therefore, to drop all thoughts of forming a fyftem of quadru-peds from the character of a fingle part: but if we take com-bined characters, of parts, manners, and food, we bid much fairer for producing an intelligible fyftem, which ought to be the fum of our aim.

THE fourth fection of digitated quadrupeds, confifts of thofe which are abfolutely deftitute of cutting teeth, fuch as the *Sloth* and *Armadillo*.

THE fifth fection is formed of thofe which are deftitute of teeth of every kind, fuch as the *Manis* and *Ant-eater*.

THE third and fourth orders, or divifions, are the *Pinnated* and the *Winged* Quadrupeds; the firft takes in the *Walrus* and the *Seals*, and (in conformity to preceding Writers) the *Manati*. But thofe that compofe this order are very imperfect: Their limbs ferve rather the ufe of fins than legs; and their element being for the greateft part the water, they feem as the links be-tween the quadrupeds and the cetaceous animals.

THE *Bats* again are winged quadrupeds, and form the next

---

* Such as the *Trichecus Rofmarus*, which has four diftinct grinders in every jaw, the *Phoca Urfina* and *Leonina*, the *Muftela Lutris*, and the *Sus Hydrochæris*; and particularly in the genus of *Vefpertilio*, which confifts of numbers of fpecies, many of which vary greatly in the number of their fore teeth.

gradation

gradation from this to the class of Birds; and these two orders are the only additions I can boast of adding in this Work.

So far of System; the rest of my plan comprehends numerous Synonyms of each Animal, a brief description, and as full an account of their place, manners, or uses, as could be collected from my own observations, or the information of others; from preceding Writers on the subject; from printed Voyages of the best authorities, or from living Voyagers, foreign and *English*; from different *Museums*, especially the public MUSEUM in our capital, from the Directors of which I have received every communication that their politeness and love of science could suggest.

I AM unwilling to weary my friends with a repetition of acknowlegements; but must renew my thanks to Sir JOSEPH BANKS, Bar^t. for variety of information collected from his papers, and from his magnificent Collection of Drawings; many of which are considerable ornaments to this Work, and to the GENERA OF BIRDS.

FROM the matchless collection of Animals, collected by the indefatigable industry of that public-spirited Gentleman, the late Sir ASHTON LEVER, I had every opportunity, not only of correcting the descriptions of the last edition, but of adding several Animals hitherto imperfectly known. His *Museum* was a liberal fund of inexhaustible knowlege in most branches of Natural History; which still remains an honor to his spirit, as well as a permanent credit and advantage to our country. It is now the property of Mr. PARKINSON, into whom no small portion of the zeal of the late enthusiastic and worthy owner for its improvement, seems to have transmigrated.

I am highly indebted to Doctor SHAW, of the *British Museum*, a rising Naturalist, for several valuable communications.

PREFACE.

To JOHN GIDEON LOTEN, Efq; late Governor in the *Dutch* fettlements in *India*, this book is under the greateft obligation for variety of remarks, relative to the Animals of the Iflands. To alleviate the cares of government, he amufed himfelf with cultivating our beloved ftudies, and brought home a moft numerous collection of Drawings, as elegant as faithful. Thefe have proved the bafis of two works: Mr. *Peter Brown* etched chiefly the contents of his *Illuftration* of Zoology from them; and the *Indian Zoology*, lately republifhed with confiderable improvements.

Mr. ZIMMERMAN, Profeffor of Mathematics at *Brunfwick*, has by his correfpondence, and his admirable book of *Zoologic Geography*, enabled me to fpeak with great precifion on the Animals of different climates, and to afcertain their different abodes and final limits.

I RESERVE for the laft acknowlegement, that learned Traveller and Naturalift Doctor PALLAS, who, under the patronage of a munificent Emprefs, hath pervaded almoft all parts of her extenfive dominion, and rendered familiar to us countries unvifited for centuries, and fcarcely known till elucidated by his labors. His liberal mind, far from thinking they fhould be *damnati tenebris*, has not only given the moft ample account of the regions he has vifited, but by a rare facility of communication, continues to inform and inftruct by correfpondence, in every matter in which his friends are defirous of information. In this light is owing, more than I can exprefs, increafe and accuracy to my prefent labors, and a vaft fund for future.

This work had once a chance of having been executed by his moft mafterly hand. I had the good fortune to meet with him at the *Hague* in 1766, when our friendfhip commenced. I there pro-

VOL. I.   b   pofed

PREFACE.

pofed to him the undertaking, and he accepted it with zeal. This preface will fhew his plan; but he was called away to greater and more glorious labors: the world need not be told how fully they have been accomplifhed.

I WILL now only add, that if this book has the fortune to be any ways ufeful to my countrymen, in promoting the knowledge of Natural Hiftory, my principal object will be anfwered: let it be treated with candor till fomething better appears; and when that time comes, the Writer will chearfully refign it to oblivion, the common fate of antiquated Syftems.

Thomas Pennant.

DOWNING,
DECEMBER, 1792.

M E T H O D.

Div. I. Hooeed Quadrupeds.
II. Digitated.
III. Pinnated.
IV. Winged.

Div. I. Sect. I.
Whole-hoofed.

Genus.
1. Horſe.

Sect. II. Cloven-hoofed.
11. Ox
111. Sheep
1v. Goat
v. Giraffe
v1. Antelope
v11. Deer
v111. Muſk
1x. Camel
x. Hog
x1. Rhinoceros
x11. Hippopotame
x111. Tapiir
x1v. Elephant.

Div. II. Digitated.

Sect. I. Anthropomorphous,
frugivorous.
xv. Ape
xv1. Maucauco.

Sect. II. With large canine teeth
ſeparated from the cut-
ting teeth, Six or more
cutting teeth in each
jaw. Rapacious, carni-
vorous.
xv11. Dog
xv111. Hyæna
x1x. Cat
xx. Bear
xx1. Badger
xx11. Opoſſum
xx111. Weeſel

b 2

* Their Element chiefly the Water.

SYSTEMATIC INDEX*.

VOL. I.

DIV. I. HOOFED QUADRUPEDS.

* All thoſe marked with an aſteriſk are added to this edition.

DIV. II. DIGITATED QUADRUPEDS.
SECT. I. Anthropomorphos *.

VOL. II.

DIV. II. Sect. II. Simply Digitated.

3

DIV.

DIV.

DIV. IV. WINGED.

A D D I T I O N S.

CATALOGUE

A LIST of Mr. PENNANT's WORKS,

PRINTED FOR BENJAMIN AND JOHN WHITE.

	£.	s.	d.
TOUR in SCOTLAND and VOYAGE to the HEBRIDES, 3 vol. 4to, *with* 132 *beautiful copper-plates, boards* — — —	3	13	6
JOURNEY from CHESTER to LONDON, 4to, *with* 23 *elegant copper-plates, boards* — —	1	5	0
TOUR in WALES, 2 vol. 4to, *with* 57 *copper-plates, and* MOSES GRIFFITH's Ten Supplemental Plates to the TOUR in WALES, 4to, *boards* —	2	9	6
BRITISH ZOOLOGY, 4 vol. 8vo, *an elegant edition, with* 284 *plates of* Quadrupeds, Birds, Fiſhes, and Shells, *boards* — — — —	2	8	0
GENERA of BIRDS, 4to, *with* 16 *plates,* and Indexes to the Ornithologie of the Comte de Buffon, and Planches Enluminées, ſyſtematically diſpoſed, *boards*	0	15	0
Index to the Ornithologie of the Comte de Buffon, and the Planches Enluminées, ſyſtematically diſpoſed, *ſewed* — — —	0	7	6

CATALOGUE of PLATES.

In the TITLE PAGE of Vol. I. the head of the *Barbary* Antelope, No. 40. The Motto is Welch, and fignifies, WITHOUT GOD IS NOTHING: WITH GOD ENOUGH.

* Thefe numbers refer to the Syftematic Index.

5

VOL. II.

FRONTISPIECE, Sea Otter. No. 286.

This figure was cut in wood by the very ingenious Mr. *Thomas Bewick*, of *Newcaſtle upon Tyne*. His hiſtory of quadrupeds, illuſtrated with ſimilar prints, has ſuch merit as to clame the attention of every naturaliſt.

LXXVIII. Sailing

HISTORY

OF

QUADRUPEDS.

Div. I. HOOFED.

SECT. I. Whole Hoofed.
II. Cloven Hoofed.

Hoof confifting of one piece.
Six cutting teeth in each jaw.

Equus *Gefner quad.* 404. *Raii fyn. quad.*
62. Pferdt. *Klein quad.* 4.
Equus cauda undique fetofa E. ca-
ballus. *Lin. fyft.* 100. Hæft. *Faun.*
fuec. No. 47.
Equus auriculis brevibus erectis, juba
longa. *Briffon, quad.* 69.

Le Cheval. *de Buffon.* iv. 174. *tab.* I.
Br. Zool. I. 1.
Wild horfe. *Leo Afr.* 339, *Hakluyt's coll.*
voy. I. 329. *Bell's trav.* I. 225.
Zimmerman. 138. 140.
Smellie's *de Buffon.* III. 306. tab. xi *.

H. with a long flowing mane; tail covered on all parts with
long hairs.

Cultivated in moft parts of the world. The moft generous
and ufeful of quadrupeds; docile, fpirited, yet obedient: adapted
to all purpofes, the draught, the road, the chace, the race. Its
voice neighing; its arms, hoofs and teeth; its tail of the ut-

* An excellent tranflation of that celebrated author, publifhed in 1785 in nine
volumes octavo, London.

VOL I.   B   moft

moſt uſe in driving off inſects in hot weather. Subject to many diſeaſes; many from our abuſe; more from our too great care of it. Its *exuviæ* uſeful: the ſkin for collars and harneſs: the hair of the mane for wigs; of the tail for the bottoms of chairs, floor-cloths, ropes, and fiſhing-lines. *Tartars* feed on its fleſh, and drink the milk of mares; and both *Kalmucks* and *Mongals* diſtil from it a potent ſpirit.

WILD IN ASIA. The horſe is found wild about the lake *Aral*; near *Kuzneck*, in lat. 54; on the river *Tom*, in the ſouth part of *Sibiria**, and in the great *Mongalian* deſerts, and among the *Kalkas*, N. W. of *China*. The *Mongalians* call them *Takija*. They are leſs than the domeſtic kind, and of a mouſe-colour, with very thick hair, eſpe-cially in winter. They have greater heads than the tame; their foreheads are remarkably arched. They go in great herds, and will often ſurround the horſes of the *Mongals* and *Kalkas* while they are grazing, and carry them away†. They are exceſſively vigilant; a centinel placed on an eminence, gives notice to the herd of any approaching danger, by neighing aloud, when they all run off with amazing ſwiftneſs. They are often ſurprized by the *Kalmucks*, who ride in amongſt them mounted on very ſwift horſes, and kill them with broad lances. They eat the fleſh, and uſe the ſkins to lie on‡. The wild horſes are alſo taken by means of hawks, which fix on the head, and diſtreſs them ſo as to give the purſuers time to overtake them. In the interior parts of *Ceylon* is a ſmall variety of the horſe, not exceeding thirty inches in height; which is ſometimes brought to *Europe* as a rarity.

* *Bell* i. 225.   † *Du Hulde*, ii. 254.   ‡ *Bell* i. 225.

The

The horse is said to be found in a state of nature in the deserts of *Africa*, to be caught there by the *Arabs*, and eaten*.

The travellers under the conduct of *Mynheer Henry Hop*, saw abundance far north of the cape; they also met with wild asses †: but have not favored us with any remarks, or descriptions of either.

Distinction must be made between the wild horses of *Asia* above mentioned, and those in the deserts on each side of the *Don*, particularly towards the *Palus Mæotis* and the town of *Backmut*. These were the offspring of the *Russian* horses employed in the siege of *Asoph* in 1697, when, for want of forage, they were turned loose, and which have relapsed into a state of nature, and grew as wild, shy, and timid as the original savage breed. The *Cossacks* chase them, but always in the winter, by driving them into the vallies filled with snow, into which they plunge and are caught; their excessive swiftness excludes any other method of capture. They hunt them chiefly for the sake of the skins: if they catch a young one, they couple it for some months with a tame horse, and so gradually domesticate it. These are much esteemed, for they will draw twice as much as the former.

The horses of the wandering *Tartars*, carried away by the herds of the wild kind, mix and breed together. Their offspring are very distinguishable by their colors, which are composed of variety of shades of chesnut.

No horses are to be met with in any place within the *Arctic* circle, except there should be a few in the extreme part of *Nor-*

---

* *Leo Afr.* Engl. ed. 340.   † Journal Historique, 40.

way.

way. They are found in *Iceland*; originally tranfported from *Norway*, and perhaps from *Scotland*, there having been an early intercourfe between it and *Iceland.* In that ifland the horfes for labor endure all the feverity of the year abroad. I imagine they live, like the rein-deer, on mofs, as they are faid to fcrape away the fnow with their feet to * get at the ground, and obtain fubfiftence. During winter, their hair grows long and thick, which preferves them againft the cold. Towards fummer they fhed their coat, and the new one is fmooth and fleek.

Kamtfchatka is entirely deftitute of horfes, and of every domeftic animal except dogs: which, with the rein-deer, are the fubftitutes of horfes ufed by the natives. *America,* before the arrival of the *Europeans,* was in like circumftances, or rather worfe; for inftead of the dog it had only a wolfifh cur; nor do either the *Greenlanders* or *Efkimaux* make any other ufe of the rein-deer, than to fupply themfelves with its flefh for food, and its fkin for raiment. But I referve a more particular account of the adventitious animals of the new world for its intended Zoology.

Dshikketaei,	Equus hemionus, *Mongolis* Dshikke- las. Nov. com. *Petrop.* xix. 394. tab. taei dictus, defcribente P. S. Pal- vii. *Zimmerman* 666.
or wild mule.	H. of the fize and appearance of the common mule, with a large head, flat forehead, growing narrow toward the nofe, eyes of a middle fize, the irides of an obfcure afh-color. Thirty-eight teeth in all; being two in number fewer than in a

* *Horrebow,* 44. They alfo refort to the fhores, and feed on the marine plants. *Von Troil,* hift. Icel. Eng. ed. 134.

common

D.shik-ketaei, or Wild. Mule.__ N.°2.

common horfe. Ears much longer than thofe of a horfe, quite erect, lined with a thick whitifh curling coat. Neck flender, compreffed: mane upright, fhort, foft, of a greyifh color: in place of the foretop, a fhort tuft of downy hair, about an inch and three quarters long.

Body rather long, and the back very little elevated. Breaft protuberant and fharp.

Limbs long and elegant: the thighs thin, as in a mule's. Within the fore legs, an oval callus, in the hind legs none. Hoofs oblong, fmooth, black. Tail like that of a cow, flender, and for half of its length naked. The reft covered with long afh-color'd hairs.

Its winter coat grey at the tips, of a brownifh afh-color beneath; about two inches long, in foftnefs like the hair of a camel; and undulated on the back. Its fummer coat is much fhorter, of a moft elegant fmoothnefs, and in all parts marked moft beautifully with fmall vortexes. The end of the nofe white; from thence to the foretop inclining to tawny. Buttocks white, as are the infide of the limbs and belly. From the mane a blackifh teftaceous line extends along the top of the back to the tail, broadeft on the loins, and growing narrower towards the tail. The color of the upper part of the body a light yellowifh grey, growing paler towards the fides. *WINTER COLOR. SUMMER COLOR.*

Length from the tip of the nofe to the bafe of the tail, fix feet feven inches. Length of the trunk of the tail one foot four; of the hairs beyond the end eight inches. The height three feet nine. *SIZE.*

Inhabits the deferts between the rivers *Onon* and *Argun* in the moft fouthern parts of *Sibiria*, and extends over the vaft plains *PLACE.*
and

and deferts of western *Tartary*, and the celebrated fandy defert of *Gobi*, which reaches even to *India*. In *Sibiria* thefe animals are feen but in fmall numbers, as if detached from the numerous herds to the fouth of the *Ruffian* dominions. In *Tartary* they are particularly converfant about *Taricnoor*, a falt lake, at times dried up. They fhun wooded tracts and lofty fnowy mountains.

MANNERS.

They live in feparate herds, each confifting of a chief, a number of mares, and colts, in all to the number of about twenty; but feldom fo many, for commonly each male has but five, and fometimes fewer females. They copulate towards the middle or end of *Auguft*, and bring for the moft part but one at a time, which by the third year attains its full growth, form, and color. The young horfes are then driven away from their paternal herds, and keep at a diftance, till they can find mates of their own age, which have quitted their dams. Thefe animals always carry their heads horizontally; but when they take to flight, hold them upright, and erect their tail. Their neighing is deeper and louder than that of a horfe.

UNTAMEABLE.

They fight by biting and kicking, as ufual with the horfe: they are fierce and untameable; and even thofe which have been taken young, are fo intractable as not to be broken by any art which the wandering *Tartars* could ufe Yet was it poffible to bring them into fit places, and to provide all the conveniencies known in *Europe*, the tafk might be effected: but I fufpect whether the fubdued animal would retain the fwiftnefs it is fo celebrated for in its ftate of nature. It exceeds that of the An-telope; it is even proverbial: and the inhabitants of *Thibet*,

GREAT SWIFT-NESS.

from the fame of its rapid fpeed, mount on it *Chammo*, their God of FIRE. The *Mongalians* defpair of ever taking it by the chace,

chace, but lurk behind some tomb, or in some ditch, and shoot them when they come to drink, or eat the salt of the desert.

They are excessively fearful, and provident against danger. A male takes on him the care of the herd, and always is on the watch. If they see a hunter, who, by creeping along the ground, has got near them; the centinel takes a great circuit, and goes round and round him, as discovering somewhat to be apprehended. As soon as the animal is satisfied, it rejoins the herd, which sets off with great precipitation. Sometimes its curiosity costs it its life; for it approaches so near as to give the hunter an opportunity of shooting it. But it is observed, that in rainy or in stormy weather, these animals seem very dull, and less sensible of the approach of mankind.

The *Mongalians* and *Tungusi* kill them for the sake of the flesh, which they prefer to that of horses, and even to that of the wild boar, esteeming it equally nourishing and wholesome *. The skin is also used for the making of boots.

Their senses of hearing and smelling are most exquisite: so that they are approached with the utmost difficulty.

The *Mongalians* call them *Dshikketaei*, which signifies the *eared*; the *Chinese*, *Yo to tse*, or *mule*†.

In antient times the species extended far to the south. It was the *Hemionos*, or *half ass*, of ARISTOTLE ‡, found in his days in *Syria*, and which he celebrates for its amazing swiftness and its fœcundity, a breeding mule being thought a prodigy ‖; and *Pliny*, from the report of *Theophrastus*, speaks of this species being found in *Cappadocia*, but adds they were a particular kind §.

* *Du Halde*, ii. 253.   † The same.   ‡ *Hist. An.* lib. vi. c. 36.
‖ *Plinii* Hist. lib. viii. c. 44.   § The same.

The

COMMON MULES.   The domeſtic mules of preſent times are the offspring of the horſe and aſs, or aſs and horſe: are very hardy; have more the form and diſpoſition of the aſs than horſe. The fineſt are bred in *Spain*; very large ones in *Savoy*. The ſynonyms of this beaſt are the following:

MULE. Mulus. *Geſner quad.* 702. *ſyn.*
quad. 64.
Maul eſel. *Klein quad.* 6.
Le Mulet. *De Buffon*,iv. 401. xiv. 336.

Briſſon quad. 71.
Equus mulus. *Lin. ſyſt. Faun. ſuec.* No. 35. *Br. Zool.* I. 13.

3. ASS.   Aſinus. *Geſner quad.* 5. *Raii ſyn. quad.* 63.
Eſel. *Klein quad.* 6.
L'ane. *De Buffon.* iv. 377.
Equus auriculis longis flaccidis, juba brevi. *Briſſon quad.* 70.

Equus aſinus. Eq. caudæ extremitate ſe-toſa, cruce nigra ſupra. *Lin. ſyſt.* 100.
Aſna. *Faun. ſuec.* No. 35. *ed.* 1746.
Aſs. *Br. Zool.* I. ii.
Smellie's de Buffon. III. 398. tab. xii.

TAME.   **H.** with long ſlouching ears, ſhort mane, tail covered with long hairs at the end. Body uſually of an aſh color, with a black bar croſs the ſhoulders.

Patient, laborious, ſtupid, obſtinate, ſlow. Loves mild or hot climates: ſcarcely known in the cold ones. Ears ſlouch moſt towards their northernly habitations. Remarkable for their ſize and beauty in *Africa* and the Eaſt.

WILD ASS, OR KOULAN.   Onager. *Varro* de re ruſt. lib. ii. c. 6.
p. 81. *Plinii* Hiſt. Nat. lib. viii. c. 44.
Oppian Cyneg. ii. Lin. 184.

Pallas in act. acad. *Petrop.* ii. 258.
Zimmerman. 666.

The *Koulan*, or aſs in a wild ſtate, muſt be deſcribed compara-tively with the foregoing ſpecies in ſome reſpects.

6

           The

The forehead is very much arched : the ears erect, even when the animal is out of order; fharp-pointed, and lined with whitifh curling hairs: the irides of a livid brown: the lips thick; and the end of the nofe floping fteeply down to the upper lip: the noftrils large and oval.

The *Koulan* is much higher on its limbs than the tame afs, and its legs are much finer; but it again refembles it in the narrow-nefs of its cheft and body; it carries its head much higher: its fcull is of a furprizing thinnefs.

The mane is dufky, about three or four inches long, compofed of foft woolly hair, and extends quite to the fhoulders; the hairs at the end of the tail are coarfe, and about a fpan long.

The color of the hair in general is a filvery white; the upper part of the face, the fides of the neck, and body, are of a flaxen-color: the hind part of the thighs are the fame; the fore part divided from the flank by a white line, which extends round the rump to the tail: the belly and legs are alfo white: along the very top of the back, from the mane quite to the tail, runs a ftripe of bufhy waved hairs of a coffee-color, broadeft above the hind part, growing narrower again towards the tail; another of the fame color croffes it at the fhoulders (of the males only) forming a mark, fuch as diftinguifhes the tame affes: the dorfal band, and the mane, are bounded on each fide by a beautiful line of white, well defcribed by *Oppian,* who gives an admirable account of the whole.

COLOR.

Its winter coat is very fine, foft, and filky, much undulated, and likeft to the hair of the camel; greafy to the touch : and the flaxen color, during that feafon, more exquifitely bright. Its fummer coat is very fmooth, filky, and even, with exception of

WINTER COAT.

SUMMER.

C certain

certain fhaded rays, that mark the fides of the neck, pointing downwards.

SIZE.

The dimenfions of a male *Koulan* were as follow : The head was two feet long : from its fetting on to the bafe of the tail was four feet ten inches and a half : the tail, to the end of the hairs, two feet one and a half : the ears eleven inches and a half. Its height before, four feet two; behind, four feet fix. It had alfo the afinine crofs on the fhoulders; which, with its fuperior fize, and ftronger formation in all its parts, diftinguifhes it at firft fight from the female.

PLACE.

This fpecies inhabits the dry and mountainous parts of the deferts of *Great Tartary*, but not higher than lat. 48. They are migratory, and arrive in vaft troops, to feed, during the fummer, in the tracts eaft and north of lake *Aral.* About autumn they collect, in herds of hundreds, and even thoufands, and direct their courfe towards the north of *India*, to enjoy a warm retreat during winter. But *Perfia* is their moft ufual place of retirement : where they are found in the mountains of *Cafbin*, fome even at all times of the year. If we can depend on *Barboga* *, they penetrate even into the fouthern parts of *India*, to the mountains of *Malabar* and *Golconda*.

ESTEEMED AS A FOOD.

According to *Leo-Africanus* †, wild affes of an afh-color are found in the deferts of northern *Africa*. The *Arabs* take them in fnares for the fake of their flefh. If frefh killed, it is hot and unfavory : if kept two days after it is boiled, it becomes excellent meat. Thefe people, the *Tartars*, and *Romans*, agreed in their preference of this to any other food : the latter indeed

* As quoted by Dr. *Pallas.*   † 340.

chufe

chuse them young, at a period of life in which it was called *Lalisio*.

> Cum tener eft *Onager*, folaque *Lalisio* matre
> Pafcitur: hoc infans, fed breve nomen habet.
>
> MARTIAL. xiii. 97.

The epicures of *Rome* preferred thofe of *Africa* to all others [*]. The grown *onagri* were introduced among the *fpectacles* of the theatre. Their combats were preferred even to thofe of the elephants. The fame poet celebrates their performances.

> Pulcher *Onager* adeft: mitti venatio debet
> Dentis *Erythræi :* Jam removete finus [†].

I can witnefs to the fpirit and prowefs of the tame afs, which diverted me much at *les combats des animaux*—the theatre, or bear-garden of *Paris*—where I faw a fight between an afs and a dog. The laft could never feize on the long-eared beaft; which fometimes caught the dog in its mouth, fometimes flung it under its knees, and kneeled on it, till the dog fairly gave up the victory.

The manners of the *Koulan*, or wild afs, are very much the fame with thofe of the wild horfe and the *Dshikketaei*. They affemble in troops under the conduct of a leader: are very fhy, but will ftop in the midft of their courfe, even fuffer the approach of man at that inftant; but will then dart away with the rapidity of an arrow difmiffed from the bow. This *Herodotus* fpeaks to, in his account of thofe of *Mefopotamia*; and *Leo Africanus*, in that of the *African*. The *Ægyptians* derive their fine breed of tame affes from them [‡].

MANNERS.

---

[*] *Plinii* Hift. Nat. lib. viii. c. 44.

[†] See alfo *Pomponius Lætus*, lib. i.; who fays the emperor *Philip* introduced twenty *Onagri*.

[‡] *Profper Alpinus*, lib. iv. c. 6.

C 2

They

They are extremely wild. HOLY WRIT is full of allusions to their favage nature. *He fcorneth the multitude of the city, neither regardeth he the crying of the driver* *. Yet they are not untameable. The *Perfians* catch and break them for the draught: they make pits, half filled with plants to leffen the fall, and take them alive. They break, and hold them in great efteem, and fell them at a high price. The famous breed of affes in the Eaft is produced from the *Koulan* reclaimed from the favage ftate, which highly improves the breed. The *Romans* reckoned the breed of affes produced from the *Onager* and tame afs to excel all others. The *Tartars*, who kill them only for the fake of the flefh and fkins, lie in ambufh and fhoot them.

They have been at all times celebrated for their amazing fwiftnefs; for which reafon the *Hebreans* called them *Pere*; as they ftyled them *Arod* from their braying †.

Their food is the falteft plants of the deferts, fuch as the *Kalis, Atriplex, Chenopodium, &c.*; and alfo the bitter milky tribe of herbs: they alfo prefer falt-water to frefh. This is exactly conformable to the hiftory given of this animal in the book of *Job*; for the words barren land, expreffive of its dwelling, ought, according to the learned *Bochart*, to be rendered *falt* places ‡. The hunters lay in wait for them near the ponds of brackifh water, to which they refort to drink: but they are not of a thirfty nature, and feldom have recourfe to water.

Thefe animals were antiently found in the *Holy Land, Syria,* the land of *Uz* or *Arabia Deferta, Mefopotamia, Phrygia,* and

---

* *Job* xxxix. 7.   † Hierozoicon, Pars i. p. 868. 869.   ‡ The fame, 872.

Lycaonia.

*Lycaonia** . But at prefent they are entirely confined to the countries abovementioned.

CHAGRIN, a word derived from the *Tartar foghré,* is made of the fkin of thefe animals, which grows about the rump, and alfo thofe of horfes, which is equally good†. There are great manufactures of it at *Aftracan,* and in all *Perfia.* It is a miftake to fuppofe it to be naturally granulated, for its roughnefs is entirely the effect of art.

The *Perfians* ufe the bile of the wild afs as a remedy againft the dimnefs of fight: and the fame people, and the *Nogayan Tartars,* have been known to endeavour the moft infamous beftialities with it, in order to free themfelves from the diforders of the kidnies.

Zebra. *Nieremberg.* 168.
Zecora. *Ludolph. Æthiop.* I. lib. i. c. 10. II. 150.
Zebra. *Raii fyn. quad.* 64. *'lein quad.* 5.
Le Zebra, ou L'ane rayè. *Briffon quad.*

70. *De Buffon,* xii. 1. *tab.* I. II.
Equus Zebra, Eq. fafciis fufcis verficolor. *Lin. fyft.* 101. *Edw.* 222.
Wild Afs. *Kolben Cape Good Hope.* ii. 112. MUS. LEV.

H. with a fhort erect mane. That, the head, and body are ftriped downwards with lines of brown, on a pale buff ground: the legs and thighs ftriped crofs-ways. Tail like that of an afs, furnifhed with long hairs at the end. Size of a common mule.

This moft elegant of quadrupeds: inhabits from *Congo* and *Angola,* acrofs *Africa,* to *Abyffinia,* and fouthward as low as the *Cape.* Inhabit the plains, but on fight of men, run into the woods and difappear. Are gregarious, vicious, untameable, ufelefs: vaftly fwift: is called by the *Portuguefe, Burro di Matto,* or wild afs.

* *Plinii* Hift. Nat. viii. c. 44.   † *Pallas;* alfo *Tavernier,* i. 21.

Will

Will couple with the aſs. A he-aſs was brought to a female zebra kept a few years ago in *London*. The *zebra* at firſt re-fuſed any commerce with it: the aſs was then painted, to re-ſemble the exotic animal. The ſtratagem took effeĉt, and ſhe admitted its embraces; and produced a mule.

5. QUAGGA.　　Le Voy. de M. *Hop.* 40. Opeagha,　LXVI. 297. or *Quagga*, of the *Hotten-*
　　　　　　　Maſſon's Travels, in the Phil. Tranſ.　*tots*. Female Zebra? *Edw.* 223 *.

H. ſtriped like the former, on the head and neck and mane. From the withers to the middle of the flanks the ſtripes grow gradually ſhorter, leaving part of the back, loins, and ſides quite plain. The ground color of the whole upper part and ſides is bay: the belly, legs, and thighs white and free from ſpots or ſtripes. The ears ſhorter than thoſe of the *Zebra*. The feet of each are ſmall, the hoofs hard.

This animal and the *Zebra* have been confounded together, and conſidered as male and female; but in each ſpecies the ſexes agree in colors and marks, unleſs that thoſe in the male are more vivid. Sir JOSEPH BANKS enabled me firſt to ſeparate them by the remarks he communicated to me on a *Quagga* he ſaw at the Cape in 1771. They keep in vaſt herds like the *Ze-bra*, but uſually in different traĉts of country, and never mix together. They are of a thicker and ſtronger make, and from the few tryals which have been made, prove of a more docil nature. A *Quagga* caught young has been known to loſe its ſavage diſpoſition, and run to receive the careſſes of mankind; and there have been inſtances of its being broke ſo far as to draw

* The loins and lower part of the back in this are ſpotted,

in

in a team with the common horfes. It is faid to be fearlefs of the *Hyæna*, and even to attack and purfue that fierce animal: fo that it proved an excellent guard to the horfes with which it was turned out to grafs at night* Nature feemed to have defigned them for the beaft of draft or of burden for this country: and they certainly might be broke for the carriage or the faddle. They are ufed to the food which harfh dry paftures of *Africa* produce; are in no terror of wild beafts, nor are fubject to the epidemic diftemper which deftroys fo many horfes of the *European* offspring; and it may generally be obferved that both the oxen and horfes introduced into this country lofe the ftrength and powers of thofe in *Europe*.

Le Gnemel ou Huemel *Molina Chili*. 303. Equus bifulcus *Gmelin Lin*. 209.

6. HUEMEL.

H. with bifulcated hoofs. Of the fize, coat, and color of an afs. The ears erect, fhort, ftrait, pointed like that of a horfe. The head equally elegant: neck and rump finely formed.

This animal inhabits the higheft and moft inacceffible part of the *Andes*, and is therefore very difficult to be taken. Yet it muft at times defcend as Com. *Byron* faw one at *Port Defire*. It neighed like a horfe; frequently ftopped and looked at our people; then ran off at full fpeed, and ftopped and neighed again†. Its voice had nothing of the braying of an afs; neither does it refemble that animal in its internal parts: is full of mettle, and of great fwiftnefs. By its cloven hoofs forms the link, as M. *Molina* obferves, between this genus and the ruminant animals.

PLACE.

* *Sparman's* Travels I. 224. † *Hawkfworth*, Vol. I. 18.

Div.

O X.

Div. I. Sect. II. Cloven Hoofed.

* with Horns.
** without Horns.

*

II. OX.

Horns bending out laterally.

Eight cutting teeth in the lower jaw, none in the upper.

Skin along the lower side of the neck pendulous.

7. BULL.

Bos *Gefner quad.* 25. *Raii fyn. quad.* 70. *Ochs. Klein quad.* 9.
Bos cornibus levibus teretibus, furfum reflexis. *Briffon quad.* 52.
Bos Taurus. B. cornibus teretibus flexis. *Lin. fyft.* 98. *Faun. fuec.* No. 48.

Le Taureau. *De Buffon*, iv. 437. *tab.* xiv.
Zimmerman, 99.
Br. Zool. I. 15.
Auer ochs. *Ridinger* wilden Thiere: tab. 37.

O. with rounded horns, with a large space between their bases.

Still found wild in small numbers, in the marshy forests of *Poland*, the *Carpathian* mountains, and *Lithuania*, and in *Asia* about mount *Caucasus*. The *Urus*, *Bonasus*, and *Bison*, of the antients. The finest and largest tame cattle in *Holstein* and *Poland*; the smallest in *Scotland*: most useful animals, every part serviceable, the horns, hide, milk, blood, fat. More subject than other animals to the pestilence. Go nine months with young.

In a wild state, the *Bonasus* of *Aristotle*, *hist. an.* ix. c. 45. and *Pliny*, *lib.* viii. c. 15. The *Urus* of *Cæsar*, *lib.* vi. c. 28. *Gesner quad.*

quad. 143. *Et Bonasus,* p. 131. and *Bison,* 140. *Bison* and *Urus Rzaczinski Polon.* 214. 228. The *Aurochs* of the *Germans.* The *American Bison,* the next to be described, differs in no respect from this.

The *Bisontes jubati* of *Scotland* are now extinct in a wild state; but their offspring, still sufficiently savage, are still preserved in the parks of *Drumlanrig* and *Chillingham.* They retain their white color, but have lost their manes*. That worthy and amiable man, my respected friend, the late *Marmaduke Tunstal,* Esq; of *Wycliff, Yorkshire,* collected several curious particulars respecting this rare breed, which are published in 1790 in a general History of Quadrupeds, illustrated with wooden plates, cut with uncommon neatness by *Thomas Bewick,* of *Newcastle upon Tyne.* His ingenuity deserves every encouragement, as his essay is the first attempt to revive with any success that long disused art, which was first begun about the year 1448. I take the liberty of inserting here a more ample account of the *Bisontes Scotici,* extracted from p. 25 of that little elegant work.

The principal external appearances which distinguish this breed of cattle from all others, are the following:—Their color is invariably white; muzzles black; the whole of the inside of the ear, and about one third of the outside, from the tip downwards, red: the color of the ears, in the undegenerated beasts, black †; horns white, with black tips, very fine and bent upwards: some of the bulls have a thin upright mane, about an inch and an half or two inches long.

At the first appearance of any person, they set off in full gallop;

---

* *Tour Scotl.* 1772. part. I. 124. part II. 284.

† About twenty years since, there were a few with *black ears;* but the present park-keeper destroyed them;—since which period there has not been one with black ears.

and at the diftance of two or three hundred yards, make a wheel round, and come boldly up again, toffing their heads in a menacing manner: on a fudden they make a full ftop at the diftance of forty or fifty yards, looking wildly at the object of their furprize; but upon the leaft motion being made, they all again turn round and fly off with equal fpeed, but not to the fame diftance; forming a fhorter circle, and again returning with a bolder and more threatening afpect than before, they approach much nearer, probably within thirty yards; when they make another ftand, and again fly off: this they do feveral times, fhortening their diftance, and advancing nearer, till they come within ten yards, when moft people think it prudent to leave them, not chufing to provoke them farther; for there is little doubt but in two or three turns they would make an attack.

CHACE.

The mode of killing them was perhaps the only modern remains of the grandeur of ancient hunting:—On notice being given, that a wild Bull would be killed on a certain day, the inhabitants of the neighborhood came mounted, and armed with guns, &c. fometimes to the amount of an hundred horfe, and four or five hundred foot, who ftood upon walls, or got into trees, while the horfemen rode off the Bull from the reft of the herd, until he ftood at bay; when the markfman difmounted and fhot. At fome of thefe huntings twenty or thirty fhots have been fired before he was fubdued. On fuch occafions the bleeding victim grew defperately furious, from the fmarting of his wounds, and the fhouts of favage joy that were echoing from every fide: but from the number of accidents that happened, this dangerous mode has been little practifed of late years, the park-keeper alone generally fhooting them with a rifled gun, at one fhot.

When

When the cows calve, they hide their calves for a week or ten days in some sequestered situation, and go and suckle them two or three times a-day. If any person come near the calves, they clap their heads close to the ground, and lie like a hare in form, to hide themselves. This is a proof of their native wildness; and is corroborated by the following circumstance that happened to the writer of this narrative, who found a hidden calf, two days old, very lean, and very weak:—On stroking its head, it got up, pawed two or three times like an old bull, bellowed very loud, stepped back a few steps, and bolted at his legs with all its force; it then began to paw again, bellowed, stepped back, and bolted as before; but knowing its intention, and stepping aside, it missed him, fell, and was so very weak that it could not rise, though it made several efforts: But it had done enough: The whole herd were alarmed, and coming to its rescue, obliged him to retire; for the dams will allow no person to touch their calves, without attacking them with impetuous ferocity.

When any one happens to be wounded, or is grown weak and feeble through age or sickness, the rest of the herd set upon it, and gore it to death.

The weight of the oxen is generally from forty to fifty stone the four quarters; the cows about thirty. The beef is finely marbled, and of excellent flavor.

Those at *Burton-Constable*, in the county of *York*, were all destroyed by a distemper a few years since. They varied slightly from those at *Chillingham*, having black ears and muzzles, and the tips of their tails of the same color; they were also much larger, many of them weighing sixty stone, probably owing to the richness

of

of the pasturage in *Holderness*, but generally attributed to the difference of kind between those with black and with red ears, the former of which they studiously endeavoured to preserve.—The breed which was at *Drumlanrig*, in *Scotland*, had also black ears.

I doubt whether any wild oxen of this species are found on the continent of *Africa*. We must beware of the misnomers of common travellers, especially the antient. Thus we shall find the wild ox of *Leo* to be the antelope, which we shall describe under the name of *Gnou*; and the buffaloes of *Pigafetta**, said to be found in *Congo* and *Angola*, may probably prove the species we describe in our number 9. A. With more confidence we may say, from the authority of *Flacourt*, that wild oxen are found in *Madagascar*, like the *European*, but higher on their legs. *Borneo*, according to *Beckman†*, and the mountains of *Java*, from the report of a worthy friend, yield oxen in a state of nature; but the torrid zone forbids the scrutiny into species, which would give satisfaction to an inquisitive naturalist. The varieties of domestic cattle sprung from the wild stock are very numerous; such as

A. The great *Indian* ox, of a reddish color, with short horns bending close to the neck; with a vast lump on the shoulders, very fat, and esteemed the most delicious part. This lump is accidental, and disappears in a few descents, in the breed produced between them and the common kind. This variety is also common in *Madagascar*, and of an enormous size.

---

* In *Purchas*. I. 1002.   † *Leo*. 304. *Flacourt*. 151. *Beckman*, 36.

Indian Ox. — A.

Lesser Indian Ox.___B.

B. A very fmall kind, with a lump on the fhoulders, and horns almoft upright, bending a little forward. This is the *Bos indicus* of *Linnæus*, and the *zebu* of *M. de Buffon*, xi. 423. tab. xlii. In *Surat* is a minute kind, not bigger than a great dog; which has a fierce look, and is ufed to draw children in fmall carts. The larger fpecies are the common beafts of draught in many parts of *India*, and draw the hackeries or chariots; and are kept in very high condition. Others are ufed as pads, are faddled, and go at the rate of twenty miles a day *.

C. Cattle in *Abyffinia*†, and the ifle of *Madagafcar*‡, with lumps on their backs, and horns attached only to the fkin, quite pendulous.

D. Cattle in *Adel*‖ or *Adea*, and *Madagafcar*, of a fnowy whitenefs, as large as camels, and with pendulous ears, and hunchbacks. They are called in the laft, *Boury* §.

E. white cattle, with black ears, in the ifle of *Tinian*.

F. the *lant* or *dant*, defcribed by LEO AFRICANUS, is another beaft, perhaps, to be referred to this genus. He fays it refembles an ox; but hath fmaller legs and comelier horns: that the hair is white; and fo fwift, as to be one of the rivals in fpeed with the *Barbary* horfe: The oftrich is the other. If the horfe can overtake either, it is efteemed at a thoufand ducats, or a hundred camels. The hoofs are of a jetty blacknefs: of the hide targets are made, impenetrable by a bullet; and valued at a great price.

G. Of the *European* cattle, the moft famous are thofe of *Holftein* and *Jutland*, which feeding on the rich low warm lowlands,

EUROPEAN.

* 'Terry's Voy. 155.   † *Lobo*, 70.   ‡ *Flacourt*, 151.   ‖ *Purchas*, II. 1106.   § *Flacourt*, 151.

between

between the two feas, grow to a great fize. A good cow yields from twelve to twenty-four quarts of milk in a day. Befides home confumption, about 32,000 are annually fent towards *Copenhagen*, *Hamburg*, and *Germany*. About the *Viftula* is bred the fame kind.

Podolia and the *Ruffian Ukrain*, particularly about the rivers *Bog*, *Dnieper*, and *Dniefter*, produce a fine breed; tall, large-horned, of a greyifh white-colour, with dufky heads and feet, and a dufky line along the back. The calves of thofe defigned for fale fuck a year, and are never worked, which brings them to a larger fize than their parents. They are called in *Germany*, *blue oxen*, 80 or 90,000 are driven to *Konigfberg*, *Berlin*, and *Bref-law*: the beft are fold at 100 rix-dollars apiece, or £ 20 fterling; which bring annually a return to their native country of 6,300,000 rix-dollars.

Hungary breeds the fame kind, and fends annually to *Vienna* and other parts of *Germany* about 120,000, which brings back 8,000,000 rix-dollars *.

The *Englifh* breed is derived from the foreign. Our native kind, fuch as the *Welfh* and *Scottifh* runts, are fmall, and often hornlefs. But by cultivation, many parts of *England* rival in their cattle many parts of the continent.

The antient *Gauls* ufed horns to drink out of; *in ampliffimis epulis pro proculis utuntur*, fays *Cæfar*: if according to *Pliny*, each horn held an *urna*, or four gallons, it was a goodly draught. *Gefner*, in his *Icon. Anim.* 34, fays, he faw a horn, he fuppofes of an *Urus*, hung againft a pillar in the cathedral of *Strafbourg*, which was fix feet long. Thefe were probably the horns of oxen, or caftrated beafts, which often grow to enormous fizes. The horns of wild cattle being very fhort.

* Doctor *Forfter*.

Taurus

American Bison — N.°7.

Taurus mexicanus. *Hernandez, mex.* 587. *de Laet,* 220. *Purchas's Pilgrims,* iv. 1561.

Bifon ex Florida allatus. *Raii fyn. quad.* 71. *Klein quad.* 13.

Buffalo. *Lawfon Carol.* 115. *Catefby App.* xxxvii. *du Pratz.* II. 49.

Bos bifon. B. cornibus divaricatis,

juba longiffima, dorfo gibbofo. *Lin. fyft.* 99.

Zimmerman, 548. No. 3.

Le Bifon d'Amerique. *Briffon quad.* 56. *de Buffon,* xi. 305.

Le Bœuf de Canada. *Charlevoix,* v. 193. *A ct. Zool.* Vol. I. No. 1. 2d Edition.

O. with fhort black rounded horns, with a great interval between their bafes. On the fhoulders a vaft hunch, confifting of a flefhy fubftance, much elevated. The fore-parts of the body thick and ftrong. The hind part flender and weak.

The hunch and head covered with a very long undulated fleece, divided into locks, of a dull ruft-color; this is at times fo long, as to make the fore-part of the animal of a fhapelefs appearance, and to obfcure its fenfe of feeing. During winter the whole body is cloathed in the fame manner. In fummer the hind-part of the body is naked, wrinkled, and dufky. The tail is about a foot long; at the end is a tuft of black hairs, the reft naked.

Inhabits *Mexico* and the interior parts of *North America.* Is found in great herds in the *Savannas;* fond of marfhy places; lodges amidft the high reeds: is very fierce and dangerous; but if taken young, is capable of being tamed. Will breed with the common kind. The only animal, analogous to the domeftic kinds, found by the *Europeans* on their arrival in the new world. Weighs from 1600 to 2900 weight.

Thefe animals are the fame with the *bifon* and other cattle, in a wild

PLACE.

SIZE.

wild state, and to be common to *Europe* and *America*.　For a fuller account, see *American Zoology*, No. 1.　I shall only say here, that before the arrival of the *Europeans*, the domestic cattle were entirely unknown in the new world.　They were equally strangers to *Kamtschatka*, its wild neighbor on the eastern side of *Asia*, till very lately, when they were introduced by the *Russians*; the first discoverers of that country.

Domestic cattle bear nearly each extreme of climate; enduring the heats of *Africa* and *India*; and live and breed within a small distance of the *arctic* circle, at *Quickjock*, in *Secha Lapmark*. So that Providence hath kindly ordered that cows, the most useful of quadrupeds, and corn, the great support of life, should bear the seasons of every country in which mankind can live.

8. GRUNTING.

Vacca grunniens villosa cauda equina, Sarluk. *Nov. com. Petrop.* v. 339. *Rubruquis voy. Harris coll.* I. 571.
Bos grunniens. B. cornibus teretibus extrorsum curvatis, vellere propendente, cauda undique jubata. *Lin. syst.* 99.
Zimmerman, 548, No. 2.

Le vache de Tartarie. *De Buffon,* xv. 136.
Le bœuf velu. *Le Brun voy. Moscov.* I. 120.
Bubel. *Bell's Travels,* I. 224.
Le Buffle a queue de cheval.
Pallas in act. acad. *Petrop.* I. pars. II. 332.

O. with a short head, broad nose, thick and hanging lips. Ears large, beset with coarse bristly hairs, pointed downwards, but not pendulous.　Horns short*, slender, rounded, up-

On the authority of Mr. *Bogle*, a most ingenious and observant traveller, who of late years penetrated from *India* into *Thibet*. See *Phil. Trans.* LXVIII. 465.

right,

Grunting Ox. — N.º 8.

right, and bending, and very sharp-pointed. They are placed remote at their bases, between which the hair forms a long curling tuft. The hair in the middle of the forehead radiated.

The space between the shoulders much elevated. Along the neck is a sort of mane, which in some extends along the top of the back to the tail. The whole body, especially the lower parts, the throat, and neck, are covered with hairs, so long as to conceal at least half the legs, and make them appear very short. All the other parts of the body are covered with long hairs like those of a he-goat. The hoofs are large: the false hoofs project much; are convex without, concave within.

Its most obvious specific mark is the tail, which, in the words of Mr. *Bogle*, spreads out broad and long, with flowing hairs like that of a beautiful mare, of a most elegant silky texture, and of a glossy silvery-color. There is one preserved in the *British Museum*, not less than six feet long. **TAIL.**

The color of the head and body is usually black; but that of the mane of the same color with the tail. **COLOR.**

Doctor *Pallas* compares the size of those which he saw to that of a small domestic cow. But the growth of these was probably checked by being brought very young from their native country into *Sibiria*. Mr. *Bogle* speaks of them as larger than the common *Thibet* breed. *Marco Polo* * says, that the wild kind, which he saw on his travels, were nearly as large as elephants. **SIZE.**

* GUILLAUME *de Rubruquis*, a friar sent by *Louis* IX. or *St. Louis*, ambassador to the *Khan* of *Tartary*, in 1253, wrote his extensive travels, and addressed them to his master. See *Purchas*, III. i. 22. MARCO *Polo* was a *Venetian* gentleman, who, in the same century, also visited *Tartary* and many other distant countries. *Purchas*, III. 65. 79.

He may exaggerate; but the tail in the *British Museum* is a proof of their great fize, for it is fix feet long, yet probably did not touch the ground; for all the figures of the animal which I have feen, do not make that part defcend quite to the heels.

Thefe animals, in the time of *Rubruquis* and *Marco Polo*, were very frequent in the country of *Tangut*, the prefent feat of the *Mongol Tartars*. They were found both wild and domefticated. They are in thefe days more rare, but are met with in abundance (I believe) in both ftates, in the kingdom of *Thibet*. Even when

fubjugated, they retain their fierce nature, and are particularly irritated at the fight of red or any gay colors. Their rifing anger is perceived by the fhaking their bodies, raifing and moving their tails, and the menacing looks of their eyes. Their attacks are fo fudden and fo rapid, that it is very difficult to avoid them. The wild breed, which is called *Bucha*, is very tremendous: if, in the chace, they are not flain on the fpot, they grow fo furious from the wound, they will purfue the affailant; and if they over-take him, they never defift toffing him on their horns into the air, as long as life remains *. They will copulate with domeftic cows. In the time of *Marco Polo*, this half-breed was ufed for the plough, and for bearing of burdens †, being more tractable than the others: but even the genuine breed were fo far tamed as to draw the waggons of the *Nomades* or wandering *Tartars*. To prevent mifchief, the owners always cut off the fharp points of the horns. The tamed kinds vary in color to red and black, and fome have horns white as ivory ‡.

---

* *Gmelin* in n. com. *Petrop.* v. 331.   † *Purchas*, III. 79.   ‡ *Witfen,*
as quoted by Dr. *Pallas.*

There

There are two varieties of the domesticated kinds, one called in the *Mongol* language *Ghainoûk*, the other *Sarlyk*. The first of the original *Thibet* race, the other a degenerated kind. Many are also destitute of horns, but have on the front, in their place, such a thickness of bone, that it is with the utmost difficulty that the persons employed to kill them, can knock them down with repeated blows of the ax *.

Their voice is very singular, being like the grunting of the hog.

A Bezoar † is said to be sometimes found in their stomachs, in high esteem among the oriental nations: but the most valuable part of them is the tail, which forms one of the four great articles of commerce in *Thibet*. They are sold at a high price, and are mounted on silver handles, and used as *chowras* or brushes to chase away the flies. In *India* no man of fashion ever goes out, or sits in form at home, without two *chowrawbadars* or brushers attending him, each furnished with an instrument of this kind ‡. The tails are also fastened by way of ornament to the ears of elephants ||, and the *Chinese* dye the hair red, and form it into tufts, to adorn their summer bonnets. Frequent mention is made of these animals in the sacred books of the *Mongols*: the cow being with them an object of worship, as it is with most of the orientalists.

Of the antients, *Ælian* is the only one who takes notice of this singular species. Amidst his immense farrago of fables, he gives a very good account of it, under the name of " the *Poe-* " *phagus*, an *Indian* animal larger than a horse, with a most thick

* *Pallas.* † *Whitsen*, as quoted by Dr. *Pallas.* ‡ Mr. *Bogle.*
|| *Bernier, Voy. Kachemire.* 124.

E 2 " tail,

" tail, and black, composed of hairs finer than the human.
" Highly valued by the Indian ladies for ornamenting their
" heads; each hair he says was two cubits long. It was the
" most fearful of animals and very swift. When it was chaced.
" by men or dogs, and found itself nearly overtaken, it would
" face its pursuers, and hide its hind parts in some bush, and
" wait for them: imagining that if · it could conceal its tail,
" which was the object they were in search of, that it would
" escape unhurt. The hunters shot at it with poisoned arrows,
" and when they had slain the animal, took only the tail and
" hide, making no use of the flesh *."

9. BUFFALO.

βοες αγριοι εν Αραχωτοις. *Arist. hist. lib.* ii. c. i.
Bos Indicus. *Plin. lib.* viii. c. 45.
Bubalus. *Gesner quad.* 122. *Raii syn. quad.* 72. *Klein quad.* 10.
Taur. elephantes *Ludolph. Æthiop.* I. lib. i. c. 10. II. 145.
Buffalo. *Dellon voy.* 82. *Faunul. Sinens.*

Bos cornibus compressis, sursum reflexis, resupinatis, fronte crispa. *Brisson quad.* 54.
Bos cornibus resupinatis intortis, antice planis. *Lin. syst.* 99.
Zimmerman. 369.
Le Buffle. *De Buffon* xi. 284. *tab.* xxv.
Br. Mus. Ashm. Mus. LEV. MUS.

O. with large horns, straight for a great length from their base, then bending upwards; not round, but compressed, and one side sharp. Skin almost naked, and black. Those about the cape of *Good Hope* of a dusky red. The head is proportionably lesser than the common ox; the ears larger: nose broad and square: eyes white: no dewlaps. The limbs long; body square; tail shorter, and more slender than that of our common cattle.

* *Ælian* de an. lib. xvi. c. xi. p. 329.

It

It grows to a very great size, if we may form a judgment from the horns. In the *British Museum* is a pair six feet six inches and a half long, it weighs twenty-one pounds, and the hollow will contain five quarts. *Lobo* mentions some in *Abyssinia*, which would hold ten. *Dellon* saw some in *India* ten feet long. They are sometimes wrinkled, but often smooth.

These animals are found wild in *Malabar*, *Borneo*, and *Ceylon* *. They are excessively fierce and dangerous if attacked: they fear fire; and are greatly provoked at the sight of red. They are very fond of wallowing in the mud; love the sides of rivers; and swim very well.

They are domesticated in *Africa*, *India*, and *Italy*, and are used for their milk and their flesh, which is far inferior to the common beef: much cheese is also produced from the milk. The horns are much esteemed in manufactories; and of their skin is made an impenetrable buff.

They form a distinct race from the common cattle. They will not copulate together, neither will the female buffaloes suffer a common calf to suck them; nor will the domestic cow permit the same from the young buffalo. A buffalo goes twelve months with young; our cows only nine†.

The buffaloes of *Abyssinia* grow to twice the size of our largest oxen, and are called *taur-elephantes*, not only on that account, but because their skins are naked and black like that of the elephant.

They are very common in *Italy*, originally introduced into *Lombardy* from *India* by king *Agilulf*, who reigned from 591 to

---

* *Dellon.* 82. *Beckman.* 36. *Knox.* 21.   † Journal historique, &c. 39.
                                                                        616.

616 *.   They are faid to have grown wild in *Apuglia*, and to be very common, in hot weather, on the fea-fhore between *Manfredonia* and *Barletta*.

The tamed kind are ufed in *Italy* for the dairy and the draught. In *India* and *Africa* for both; and in fome parts of *India* alfo for the faddle.

Ariftotle defcribes thefe animals very well under the title of wild oxen, among the *Arachotæ*, in the northern part of *India*, bordering on *Perfia*. He gives them great ftrength, a black color, and their horns bending upwards more than thofe of the common kind. *Pliny* probably means a large breed of this kind, as high as a camel, with horns extending four feet between tip and tip.

A. Naked: a fmall fort, exhibited in *London* fome years ago, under the name of *Bonafus*; of the fize of a runt: hair on the body briftly, and very thin, fo that the fkin appeared: the rump and thighs quite bare: the firft marked on each fide with two dufky ftripes pointing downward, the laft with two tranfverfe ftripes: horns compreffed fideways, taper, fharp at the point. *Eaft Indies*.

B. The *Anoa* is a very fmall fpecies of buffalo, of the fize of a middling fheep. They are wild, in fmall herds, in the mountains of *Celebes*, which are full of caverns. Are taken with great difficulty; and even in confinement are fo fierce, that Mr. *Soten* loft in one night fourteen ftags, which were kept in the fame paddock, whofe bellies they ripped up.

* Tunc primum caballi fylvatici et *Bubali* in *Italiam* delati, *Italia* populis miraculo fuerunt. *Warnefridi de geftis Longobarder*. Lib. iv. c. ii. *Miffon's voy*. iv. 392.

C. The

Naked Buffalo.——A.

C. The *Gauvera* is a species of ox found in *Ceylon*, and described by *Knox*, p. 21; who says, its back stands up in a sharp
ridge, and whose legs are white half way from the hoofs. I have
received an account of hunch-backed oxen being found in that
island, which are probably the animals intended by Mr. *Knox*.

Le Bœuf Musquè. de *M. Jeremie, Voy-*   v. 194. *Arƈt. Zool.* vol. I. No. 2.   10. MUSK.
ages au Nord. iii. 314. *Charlevoix.*   LEV. MUS..

O. with horns very closely united at the base, bending in-   BULL.
wards and downwards, and turning outwards at their
points; two feet round at the base, and vastly prominent, rising
just on the top of the forehead; length only two feet; very
sharp at the points: head and body universally covered with very
long silky hairs, of a-dark color: some of the hairs are seventeen inches long. Beneath them, in all parts, in great plenty, and
often in flocks, is a cinereous wool of exquisite fineness. M. *Jeremie*
brought some to *France*, of which stockings were made more beautiful than those of silk. The tail is only three inches long, a mere   TAIL.
stump, covered with very long hairs.

The horns of the cow are nine inches distant from each other   THE COW
at the base, and are placed exactly on the sides of the head;
are thirteen inches long, and eight inches and a half round at the
base. The flesh scents strong of musk: the length of the skin
of the cow was six feet four inches; including the head, which was
fourteen inches long: the legs very short: the hair trails on the
ground; so that the whole animal seems a shapeless mass, without distinction of head or tail: the shoulders rise into a lump.
In size lower than a deer.

This

This animal is very local: it appears firſt between *Churchill* river and thoſe of *Seals* on the weſtern ſide of *Hudſon's Bay*: are very numerous between *Lat.* 66. and 73 North; and go in herds of twenty and thirty: delight in barren and rocky mountains: and run nimbly, and are very active in climbing the rocks: ſeldom frequent the woody parts: are ſhot by the Indians for the ſake of the ſkins, which make the beſt and warmeſt blankets.

They are found again among the *Cris*, or *Criſtinaux*, and the *Aſſinibouels*, and among the *Attimoſpiquay*: are continued from theſe countries ſouthward as low as the provinces of *Nievera* and *Libola* · for Father *Marco di Nica* and *Gomara* plainly deſcribe both kinds *.

A part of this ſpecies has been found in the north of *Aſia*, the head of one having been diſcovered in *Sibiria*, on the arctic moſſy flats, near the mouth of the *Oby*. It is to Doctor *Pallas* † I owe the account; who does not ſpeak of this kind as being foſſil, but ſuſpects that the whole carcaſe was brought on floating ice from *America*, and depoſited where the ſcull was found. If this is certain, it proves that theſe animals ſpread quite acroſs the continent of *America* from *Hudſon's Bay* to the *Aſiatic* ſeas.

11. CAPE.   O. with the face covered with long harſh black hair. Chin, underſide of the neck, and dewlap, covered with long, pendulous, and coarſe hairs of the ſame color. From the horns, along the top of the neck, to the middle of the back, is a long looſe black mane. Body covered with ſhort, dark, cinereous

* *Purchas,* iv. 1561, v. 854.   † Nov. com. xvii. 601. tab. 17.

3   hair:

hai : bafe of the tail almoft naked and cinereous, the reft full of long black hair. Skin thick and tough.

Horns * thick at the bafe, bend outwards, then fuddenly invert. Length along the curve one foot nine: from tip to tip eight inches and a half. Between each at the bafe three inches. The horns, *tab.* fig. iii. p. 9. of my former edition, which I attributed to the next fpecies, moft probably are thofe of a young animal of this kind. They are defcribed by *Grew*, p. 26. of his account of the Mufeum of the Royal Society; but he improperly thinks them the horns of the common *buffalo*.

HORNS.

Length from nofe to tail, of one not of the largeft fize, is eight feet: the height five and a half. Depth of the body three feet: length of the head one foot nine: of the trunk of the tail one foot nine: to the end of the hairs, two feet nine. Body and limbs thick and ftrong. Fore legs two feet and a half long.

The face is covered with black coarfe hairs. From the chin along the throat and dewlap was a quantity of very long pendulous hairs, and from the hind part of the horns, ran on the middle of the back a long loofe black mane. The body was covered with fhort dark cinereous hair. The bafe of the tail almoft naked: the reft full of long black hair. In aged bulls the hair is of a deep brown color, about an inch long, and very thin†. The former I defcribed from a very entire fkin, brought from the *Cape* by Sir JOSEPH BANKS. It agreed in all the meafurements with a bull of this fpecies killed by Doctor SPARMAN in his *African* expedition, excepting in the horns: it poffibly might have

HAIR.

* M. de *Buffon* has engraven the horns, vol. xi. 416. tab. xli.
† *Sparman's* travels. II. 64. tab. II.

VOL. I.   F   been

OTHER HORNS. been the skin of a younger animal, or of a female. Those described by Mr. *Sparman* occupied at their bases a circumference of about eighteen or twenty inches, and were placed about an inch distant from each other. Their upper surface was much elevated and very rugged, with hollows an inch deep. They spread far over the head towards the eyes, then grew taper and bent down on each side of the neck; and the ends inclined backwards and upwards. The space between the point sometimes is not less than five feet. The weight of a pair in the *Leverian* Museum was twenty five pounds. The ears are a foot long, and swag in a pendulous manner beneath the bottom of the horns.

PLACE.   They inhabit the interior parts of *Africa*, north of the Cape of *Good Hope*; but, I believe, do not extend to the north of the Tropic. They are greatly superior in size to the largest *English* ox: hang their heads down, and have a most fierce and malevolent appearance, which is increased by a method they have of holding their heads aside, and looking askance with their eyes sunk beneath FIERCENESS AND the prominent orbits: are excessively fierce and dangerous to CRUELTY. travellers: will lie quietly in wait in the woods, and rush suddenly on passengers, and trample them, their horses, and oxen of draught, under their feet *: so that they are to be shunned as the most cruel beasts of this country. They are not content with the death of man or animal which have fallen in their way; but they will return to the slaughtered bodies as if to satiate their revenge, stand over them for a time, trample on them, crush them with their knees, and with horns and teeth deliberately mangle the whole body; repeating this species of insult at certain intervals, and with their rough tongues entirely strip off by licking, the skin from

* *Forster*'s Voy. i. 83, *Masson*'s Travels. Phil. Transf. lxvi. 296.

corps,

corps, exactly in the manner in which *Oppian* informs us that the *Thracian* bisons did the slain in times of old. They are prodigiously swift, and so strong, that a young one of three years of age, being placed with six tame oxen in a waggon, could not by their united force be moved from the spot.

They are also found in the interior parts of *Guinea* *; but are so fierce and dangerous, that the negroes who are in chace of other animals are fearful of shooting at them. The lion, which can break the back of the strongest domestic oxen at one blow, cannot kill this species, except by leaping on its back, and suffocating it, by fixing its talons about its nose and mouth †. The lion often perishes in the attempt; but leaves the marks of its fury about the mouth and nose of the beast. It loves much to roll in the mud, and is fond of the water.

The flesh is coarse, but juicy, and has the flavor of venison; and the marrow most delicate. The bones are of most uncommon strength and hardness. The animals are shot with balls of the weight of two ounces and a quarter, and hardened by an alloy of tin, yet are usually flattened or shivered to pieces when they happen to strike against a bone.

The hides are thick and tough, and of the first use among the *African* colonists for the making of thongs, halters, and harnesses. On them alone they depend on security of their horses or oxen, which, on the approach of a lion or other wild beast, would snap any other in their efforts to get loose.

They live in great herds, even of thousands, especially in *Krake-*

* Mr. *Smeathman*, a gentleman long resident in *Guinea* on philosophical researches.

† *Sparman, Stock. Welt. Handl.* 1779, p. 79. tab. iii. and Travels, II. 63.

Kamma, and other deferts of the *Cape*; and retire during day into the thick forefts. They are called by the *Hottentots t'Kau*, by the *Dutch* of the Cape *Aurochs*, but differ totally from the *European*.

Another fpecies of *Aurochs* is briefly defcribed by the *Dutch* travellers *; who fay it is like the common ox, but larger, and of a grey color; that its head is fmall, and horns fhort; that the hairs on the breaft are curled; that it has a beard like a goat; and that it is fo fwift, that the *Namacques* call it *Baas*, or the *Mafter-courier*. They diftinguifh this from the *Gnou*, No. 16 of this work, or I fhould think it the fame animal.

12. DWARF.	Un moult beau petit bœuf d'Afrique, ribus, dorfo gibbo juba, nulla. *Lin.* *Belon voy.* 119, 120. *fyft.* 99. Bos Indicus. B. cornibus aure brevio- *Zimmerman*, 459. No. 6.

O. with horns receding in the middle, almoft meeting at the points, and ftanding erect: in body larger than a roe-buck, lefs than a ftag: compact and well made in all its limbs: hair fhining, of a tawny brown: legs fhort, neck thick, fhoulders a little elevated: tail terminated with long hairs, twice as coarfe as thofe of a horfe.

This fpecies is defcribed by *Belon*, who met with it at *Cairo*; but he fays, that it was brought from *Afamie*, the prefent *Azafi*, a province of *Morocco*, feated on the ocean. I fufpect it to be the *lant*, mentioned, p. 17, which may vary in color.

* Journal hiftorique, 43. 46.

Horns

Horns twifted fpirally, and pointing outwards.

Eight cutting teeth in the lower jaw, none in the upper.

Ovis. *Plinii. lib.* viii. c. 47. *Gefner quad.* 771. *Raii fyn. quad.* 73.
Widder Schaaf- *Klein quad.* 13.
La brebis. *de Buffon,* v. 1. *tab.* I. II.
Aries Laniger cauda rotunda brevi. *Brif-* *fon quad.* 48.
Ovis aries. O. cornibus compreffis lunatis. *Lin. fyft.* 97.
Far. *Faun. fuec.* No. 45.
Zimmerman. 112.

13. COMMON.

THE fheep, the moft ufeful of the leffer animals; the fource of wealth in civilized nations. *England,* once the envy of *Europe* for its vaft commerce in the produ&s of this creature, now begins to be rivalled by others, thro' the negle&, the luxury, the too great avidity of our manufa&turers. The *Englifh* wool excellent for almoft every purpofe. The *Spanifh* extremely fine; the œconomy of the fhepherds admirable; as is their vaft attention to the bufinefs, and their annual migrations with their flocks. The fineft fleeces in the world are thofe of *Caramania**, referved entirely for the *Moulbaes* and priefts; thofe of *Cachemire* † excellent; and the Lamb-fkins of *Bucharia* exquifite ‡.

The fheep in its nature harmlefs and timid: refifts by butting

* *Chardin's Travels* in *Harris's Coll.* ii. 878. and *Tavernier,* i. 40.

† *Bernier's Voy.* ii. 94.

‡ *Bell's Travels,* i. 46. Thefe fkins bear a great price, have a fine glofs, and rich look.

with

with its horns: threatens by ftamping with its foot: drinks little: generally brings one at a time, fometimes two, rarely three: goes about five months with young: is fubject to the rot; worms in its liver; the vertigo.

A. COMMON Sh.

With large horns, twifting fpirally and outwardly.

Ovis ruſtica. *Lin. ſyſt.* 97. *Zimmerman.* 112. LEV. MUS.

Sheep have their teeth, when they feed in certain paftures, incrufted and gilt with pyritical matter; which has been obferved in the fheep of *Ægypt*, *Anti-Lebanon*, and *Scotland* *.   I never faw an inftance of it in thofe animals: but have met with the teeth of oxen, in the *Blair* of *Athol*, *N. Britain*, covered with a gold-colored fubftance.

B. CRETAN Sh. Ovis Strepſiceros. La Chevre de *Crete*. *Briſſon quad.* 48. *Raii ſyn. quad.* 75.   Cornibus rectis *Zimmerman.* 131. carinatis flexuofo-ſpiralibus. *Lin.ſyſt.* Strepſicheros ou Mouton de *Crete*. *Belon* 98.   *voy.* 16. *Geſner quad.* 308. *Icon.* 15.

Has large horns, quite erect, and twifted like a fcrew; common in *Hungary*.   Is called by the *Auſtrians*, *Zackl*; and is almoft

* *Haſſelquiſt's Trav.* 192. *Sib. Scot.* lib. iii. 8.

the

the only kind which the butchers deal in·*. Great flocks are found on Mount *Ida* in *Crete*. *De Buffon* has given figures of a ram and ewe, under the name of *Vallachian* Sheep †.

C. Hornless. Ovis Anglica. *Lin. syst.* 97.

Common in many parts of *England*; the largest in *Lincolnshire*, the left horned sheep in *Wales*.

D. Many-horned. Ovis polycerata. *suppl.* iii. *p.* 73. *Zimmerman*, 127. 128. *Lin. syst.* 97. *de Buffon* xi. *tab.* xxxi. Lev. Mus.

Common in *Iceland*, and other parts of the *North*; they have usually three horns, sometimes four, and even five. Many-horned sheep are also very common in *Sibiria*, among the *Tartarian* flocks, about the river *Jenesei* ‡. The horns of these grow very irregularly, and form a variety totally different from the next.

E. I have engraven a very singular ram, with two upright and two lateral horns: body covered with wool: fore part of the neck with yellowish hairs, 14 inches in length: was alive in *London* a few years ago: very mischievous and pugnacious: the horns the same with those in *Grew, tab.* ii. M. *De Buffon* has engraven one of the same kind, but with only two horns, under the name of *Le Morvant de la Chine* ||. The animal which I saw

* *Kramer* anim. *Austriæ*, 322. † Suppl. iii. 66. tab. vii. viii. ‡ *Pallas* Spicil. Zool. fasc. xi. 71. tab. iv. & v. || Suppl. iii. 68. tab. x.

was

was brought from *Spain*; but I am uncertain whether it was a native of that country.

F. A moſt elegant ſpecies, brought from *Guinea*, and preſented to me by *Richard Wilding*, Eſq; of *Llanrhaidr*, in *Denbighſhire*. It was ſmall of ſtature, and moſt beautifully limbed. The hair of a ſilvery whiteneſs, and quite ſilky; on the fore and hind part of the neck of a great length, eſpecially in front; half of the noſe was of a jetty blackneſs; on each knee and on each ham was a black ſpot; the footlock and feet black. It had only two horns.

In the month of *November* it began to aſſume a ſoft woolly coat, like that of the *Engliſh* ſheep: ſo ſenſibly was it influenced by climate. When I firſt received this animal it was extremely gentle; attended, me like a dog, in all my walks; and leaped over every ſtile in its way. It afterwards (on being introduced to ſome females) grew ſo vicious as to become dangerous, ſo I was obliged to ſend it to a mountain-incloſure, where it died.

G. AFRICAN. Aries guineenſis. *Mar-grave Braſil.* 134. *Raii ſyn. quad.* 75.
Le Belier des Indes. *de Buffon.* xi. 362. tab. xxxiv. &c.
Ovis guineenſis, O. auribus pendulis, palearibus laxis piloſis. *Lin. ſyſt.* 98. *Zimmerman.* 131.
La Brebis de Guinee. *Briſſon quad.* 51.
Sheep of *Sahara. Shaw's travels*, 241.
Carnero or Bell wether. *Della Valle trav.* 91.

Meagre; very long legged and tall: ſhort horns: pendent ears, covered with hair inſtead of wool: ſhort hair: wattles on the neck. Perhaps the *Adimain* of *Leo Africanus*, 341; which he ſays furniſhes the *Lybians* with milk and cheeſe; is of the ſize of an aſs, ſhape of a ram, with pendent ears. *Della Valle* tells us, that at *Goa* he has ſeen a wether bridled and ſaddled, which car-ried

E. *Four horned Ram.* — D. *Horns of the Iceland Sheep.*
B. *Horns of the Cretan Sheep.*

ried a boy twelve years old. The *Portuguese* call them *Cabritto*. They are very bad eating.

H. **BROAD-TAILED.** *Ludolph. Æthiop.* 53. Ovis arabica. *Caii opusc.* 72. *Gesner quad. Icon.* 15. *Faunul. Sinens.* Ovis laticauda. *Raii syn. quad.* 74. *Zimmerman.* 129. *Lin. syst.* 97. *Brisson quad.* 50. *Nov. Com. Petrop.* v. 347. *tab.* viii. Le Mouton de Barbarie. *de Buffon*, xi. 355. *tab.* xxxiii. *Shaw's travels*, 241. *Russel's Aleppo*, 51.

Common in *Syria, Barbary,* and *Æthiopia.* Some of their tails end in a point, but oftener square or round. They are so long as to trail on the ground, and the shepherds are obliged to put boards with small wheels under the tails to keep them from galling. These tails are esteemed a great delicacy, are of a substance between fat and marrow, and are eaten with the lean of the mutton. Some of these tails weigh 50 lb. each.

The short thick-tailed sheep are common among the *Tartars* *.

The broad-tailed sheep are found in the kingdom of *Thibet*; and their fleeces, in fineness, beauty, and length, are equal even to those of *Caramania.* The *Cachemirians* engross this article, and have factors in all parts of *Thibet* for buying up the wool, which is sent into *Cachemir*, and worked into *shauls*, superior in elegance to those woven even from the fleeces of their own country. This manufacture is a considerable source of wealth†. *Bernier* relates, that in his days, *shauls* made expresly for the great *omrahs*, of the *Thibetian* wool, cost a hundred and fifty rou-

---

* *Pallas* Spicil. Zool. fasc. xi. tab. iv. fig. 2. a.

† Phil. Transf. lxvii. 485. From Mr. *Bogle's* account.

pees: whereas thofe made of the wool of the country never coft more than fifty *.

Thefe articles of luxury have, till of late, been fuppofed to have been made with the hair of a goat, till we were undeceived by Mr. *Bogle*, a gentleman fent by Mr. *Haftings* on a commiffion to the *Tayfhoo Lama* of *Thibet*. His account of that diftant country is inftructive and entertaining. We have fufficient in the Philofophical Tranfactions to make us regret that we have not the whole of that memorable miffion.

Both the broad-tailed and long-tailed varieties were known to the antients. The *Syrian* are the kind mentioned. *Ariftotle* takes notice of the firft, *Pliny* of the fecond. One fays the tails were a cubit broad; the other, a cubit in length †.

I. The fat-rumped fheep; without tails: arched nofes; wattles; pendulous ears; and with curled horns, like the common fheep. The wool coarfe, long, and in flocks: legs flender: head black. Ears of the fame color, with a bed of white in the middle. The wool is generally white; fometimes black, reddifh, and often fpotted.

The buttocks appear like two hemifpheres, quite naked and fmooth, with the *os coccygis* between fcarcely fenfible to the touch. Thefe are compofed only of fuet; whence Dr. *Pallas* properly ftyles this variety *ovis fteatopyges*. Thefe fheep grow very large, even to two hundred pounds weight, of which the pofteriors weigh forty.

* *Bernier's* voy. *Cachemir.* 95. By miftake he calls it the hair of a goat from. *Great Thibet.*

† *Arift.* hift. an. viii. c. 28. *Plin.* viii. c. 48.

Their

Cape Sheep.

Their bleating is short and deep, more like that of a calf than sheep.

They abound in all the deserts of *Tartary*, from the *Volga* to the *Irtis*, and the *Altaic* chain: but are more or less fat according to the nature of the pasture: but most so where the vernal plants are found; and in the summer, where there are herbs replete with juice and salts, and where salt springs and lakes impregnate the vegetation of the country. These monstrous varieties are supposed to originate from disease, arising from an excess of fat in the hind parts, which involved*, and at length destroyed the tail.

By breeding between animals similarly affected, the breed was continued in those parts where food and climate have concurred to support the same appearances. Those with fat tails, mentioned in the variety G, are rather in the way to exhibit such singularity as this variety, or are a mixed breed between the common and the tail-less kind.

All abound so greatly in *Tartary*, that 150,000 have been sold annually at the *Orenburg* fairs, and a much greater number at *Troinkaja*, in the *Irkutsk* government, bought from the *Kirgisian Tartars*, and dispersed through *Russia*. They are very prolific: usually bring two at a time, often three.

The next to be taken notice of is the stock from which the whole domestic race is derived.

---

* This is exemplified in fig. 1. tab. iv. *Zimmerman.* 132.

Musimon,

S H E E P.

H. Wild*.

Mufimon, *Plinii lib.* viii. *c.* 49.
Ophion, *lib.* xxviii. *c.* 9. xxx. *c.* 15.
Mufmon feu Mufimon, *Gefner quad.* 823.
  Zimmerman. 114. 546.
Capra Ammon, *Lin. fyfl.* 97.
Le Chamois de Siberie, *Briffon quad.* 42.

& la chevre du Levant, 46.
Le Mouflon, *de Buffon*, xi. 352. *tab.* xxix.
Rupicapra cornubus arietinis. Argali, *Nov. com. Petrop.* iv. 49. 388. *tab* viii.

SIBIRIAN.

1. Sh. with horns placed on the fummit of the head, clofe at their bafes, rifing at firft upright, then bending down and twifting outward, like thofe of the common ram; angular, wrinkled tranfverfely. In the FEMALES leffer and more upright, and bending backwards.

Head like a ram; ears leffer than in that animal; neck flender; body large; limbs flender but ftrong; tail very little more than three inches long: hoofs fmall, and like thofe of a fheep.

SUMMER COAT.

Hair in the fummer very fhort and fmooth, like that of a ftag: the head grey: the neck and body brownifh, mixed with afh-colour: at the back of the neck, and behind each fhoulder, a dufky fpot: fpace about the tail yellowifh.

WINTER.

In the winter, the end of the nofe is white; face cinereous; back ferruginous, mixed with grey, growing yellowifh towards the rump: the rump, tail, and belly white: the coat in this feafon rough, waved, and a little curling; an inch and a half long; about the neck two inches; and beneath the throat ftill longer.

* It is called by the *Kirgifian Tartars, Argali,* perhaps from *Arga,* an Alpine fummit: the ram, *Guldfha.* By the *Kamtfchatkans, Goâdinachtfch;* and by the *Kuritians, Rikun-donotoh,* or the *Upper Rein Deer,* from its inhabiting the loftier parts of the mountains. The *Ruffians* ftyle it *Stepnoi Barann,* or the *Ram of the Defert; Kamennoi,* or the *Rock Ram,* and *Dikoi,* or the *wild.* PALLAS.

The

The ufual fize of the male is that of a fmaller hind; the females lefs: the form ftrong and nervous.

2. The fecond animal which I defcribe related to this fpecies, is the Μουσμον of *Strabo*, and *Mufmon* of *Pliny*; perhaps alfo the *Ophion* of the latter, and the wild ram of *Oppian**, which with its horns often laid proftrate even the wild boar: Thefe were natives of *Spain*, *Sardinia*, and *Corfica*, and are ftill exifting in thofe iflands. I have feen a pair from the firft at *Taymouth*, the feat of the Earl of *Breadalbane*, and another pair from the laft at *Shugborough*, the feat of the late *Thomas Anfon*, Efq.

The laft I defcribe thus. The height of the male, to the top of the fhoulders was two feet and a half: irides a light yellowifh hazel: horns, ten inches and a half long, five and a half round at the bafe, twelve inches diftant between tip and tip: *finus lacrymalis* very long. Ears fhort and pointed; brown and hoary without, white within. Head fhort and brown; lower part of the cheeks black; fides of the neck tawny: lower part covered with pendent hairs fix inches long, and black. Body and fhoulders covered with brown hairs, tipped with tawny: on the middle of the fides a white mark pointing from the back to the belly. Belly, rump, and legs white; the laft have a dufky line on their infides. Tail fhort: *fcrotum* (as common to all) pendulous, like that of a ram.

The remains of *Martino*, a male animal of this kind, imported from *Corfica* by the illuftrious defender of the liberties of his country, General *Paoli*, is now preferved in the *Leverian Mufeum*. It was of the age of four years at the time of its deceafe. Its

---

* Cyneg. ii. 330. Ophion *Plinii* lib. xxviii. c. 9. xxx. c. 15.

horns

horns are twenty-two inches long; the space between tip and tip near eleven; the girth near the base the same. This poor animal had the ill fortune to fall, in our land of freedom, into heavy slavery, and hard usage, in the latter part of his life, which stinted its growth, and prevented the luxuriancy of its horns; which ought, at its age, to have had the volutes of a large-horned ram, to have been fifteen inches round at the base, and to have resembled those of the painting by *Oudry*.

The colors of this specimen differed a little from the others. On the front of the neck is a large spot of white. The shoulders were covered with black hairs; bright and glossy in a state of vigor. On each side of the back, near the loins, is a large bed of white. The eyes, when in health, large, bright, and expressive.

The male, in its native country, is called *Mufro*, the female *Mufra*. They inhabit the highest parts of the *Corsican* alps, unless forced down by the snows into rather lower regions. They are so wild, and so fearful of mankind, that the old ones are never taken alive: but are shot by the *chasseurs*, who lie in wait for them.

The females bring forth in the beginning of *May*, and the young are often caught after their dam is shot. They instantly grow tame, familiar, and attach themselves to their master. They will copulate with the sheep: there is now an instance in *England* of a breed between the ram of this species, and a common ewe. They are likewise very fond of the company of goats.

In a wild state, they feed on the most acrid plants: and when tame will eat tobacco, and drink wine.

Their

Their flesh is favory, but always lean. The horns are ufed for powder-flafks, flung in a belt, by the *Corfican* peafants; and fome are large enough to hold four or five pounds, of twelve ounces each.

The *Sardinians* make ufe of the fkins dreffed, and wear them under their fkirts, under the notion of preferving them againft bad air. They alfo wear a furtout without fleeves, made of the fame materials, which falls below the knees, and wraps clofe about their bodies. The fkin is very thick, and might have been proof againft arrows, when thofe miffile weapons were in ufe. At prefent thefe furtouts are worn to defend them againft briars and thorns, in paffing through thickets. In all probability they are the very fame kind of garment as the *maftruca fardorum* *, which the commentators on *Cicero* fuppofe to have been made of the fkins of the *Mufro:* and the *Maftrucati Latrunculi* † the people who wore them. This is in a manner confirmed, as they are ftill in ufe with the *latre* or banditti of the ifland; who find the benefit of them in their impetuous fallies out of the brakes of the country, on the objects of their rapine.

MASTRUCA SAR-
DORUM.

The race is at prefent extinct in *Spain*; but is ftill found in *Sardinia* and *Corfica:* whether it exifts ftill in *Macedonia* ‡, we are ignorant. It is found in thefe days in great abundance, but confined to the north-eaft of *Afia*, beyond the lake *Baikal*, between the *Onon* and *Argun*, and on the eaft of the *Lena* to the height of

PLACE.

* Quem purpura regalis non commovit, eum *fardorum maftruca* tentavit. Oratio pro M. *Æmilio Scauro.*

† Cum *maftrucatis* latrunculis a propraetore una cohorte auxiliaria gefta, &c. De Provinciis conful.

‡ *Belon* has given, in his *Obfervations*, &c. p. 54. a figure and very accurate defcription of this animal, under the name of *Tragelaphus.* As he then wrote from Mount *Athos*, it probably was an inhabitant of the chain of mountains continued from that famous promontory.

lat.

lat. 60; and from the *Lena* to *Kamtſchatka*; and perhaps the *Kurili* iſlands. It abounds on the deſert mountains of *Mongalia, Songaria* and *Tartary*. It inhabits the mountains of *Perſia*, and the north of *Indoſtan* *. The breed once extended further weſt, even to the *Irtis*; but as population increaſed, they have retired to their preſent haunts, ſhunning thoſe of mankind.

CALIFORNIA.

It is probable that theſe animals are alſo found in *California*. The Jeſuits who viſited that country in 1697, ſay that they found a ſpecies of ſheep as big as a calf of a year or two old, with a head like that of a ſtag, and enormous horns like thoſe of a ram; and with tail and hair ſhorter than that of a ſtag. This is very likely, as the migration from *Kamtſchatka* to *America* is far from being difficult.

ONCE IN BRITAIN.

They were once inhabitants of the *Britiſh* iſles. *Boethius* mentions a ſpecies of ſheep in *St. Kilda*, larger than the biggeſt hegoat, with tails hanging to the ground, and horns longer, and as thick as thoſe of an ox †. This account, like the reſt of his hiſtory, is a mixture of truth and fable: I ſhould have been ſilent on this head, had I not better authority; for I find the figure of this animal on a *Roman* ſculpture, taken out of *Antoninus*'s wall near *Glaſgow* ‡. It accompanies a recumbent female figure, with a rota or wheel, expreſſive of a *via* or way, cut poſſibly into *Caledonia*; where theſe animals might, in that early age, have been found. Whether they were the objects of worſhip, as among the antient *Tartars*, I will not pretend to ſay; for among the graves of thoſe diſtant *Aſiatics*, brazen images and ſtone figures of their *argali*, or wild ſheep, are frequently found ‖.

---

* Dr. PALLAS.   † *Boeth.* deſc. Regn. *Scotiæ*, 8.

‡ Plates of the ſculptures, publiſhed by the univerſity of *Glaſgow*, tab. xvi.

‖ Pallas Spicil. Zool. faſc. xi. 19. *Strahlenberg*'s Hiſt. *Ruſſia*, tab. B.

Their

Their prefent habitations, in *Sibiria*, are the fummits of the higheft mountains, expofed to the fun, and free from woods. They go in fmall flocks; copulate in autumn *, and bring forth, in the middle of *March*, one, and fometimes two young. At that feafon the females feparate from the males, and educate their lambs; which when firft dropped are covered with a foft grey curling fleece, which changes into hair late in the fummer. At two months age the horns appear, are broad, and like the face of an ax. In the old rams they grow of a vaft fize. They are fometimes found of the length of two *Ruffian* yards, meafured along the fpires; weigh fifteen pounds apiece; and are fo capacious as to give fhelter to the little foxes, who find them accidentally fallen in the wildernefs. Father *Rubruquis*, the traveller of 1253, firft takes notice of thefe animals, under the name of *Artak*. He fays he had feen fome of the horns fo large, that he could fcarcely lift a pair with one hand; and that the *Tartars* made great drinking-cups with them †.

They feed from fpring to autumn in the little vallies among the tops of the mountains, on young fhoots and *Alpine* plants, and grow very fat. Towards winter they defcend lower, eat either the dry grafs, perennial plants, moffes, or lichens; and are found very lean in the fpring. They are then purged by the early *pulfatillæ*, and other fharp *anemonoid* plants, of which the tame fheep are alfo exceffively fond. They, befides, at all times of the year, frequent the places abundant in falt, as is frequent in every part of *Sibiria*, and excavate the ground, in order to get more readily at it. Thefe anfwer to the *licking-places* in *America*, and are the haunts of deer as well as *argali*.

* *Gmelin*, in Nov. Com. *Petrop*. iv. 390. † *Purchas*, iii. 6.

SHEEP.

They are very fearful of mankind: when clofely purfued, they do not run in a progreffive courfe, but obliquely from fide to fide, in which they fhew the nature of fheep. They ftrive as foon as poffible to reach the rocky mountains, which they afcend with great agility; and tread the narroweft paths over the moft dangerous precipices with the greateft fafety.

The old rams are very quarrelfome, and have fierce combats among themfelves, fighting with their heads, like the common kind. They often ftrike their antagonift down the fteep precipices; and their horns and bones are frequently found at the bottom; a mark of the fatal effects of their feuds. They will often entangle their horns accidentally, and thus locked, fall down together and perifh.

Uses.

They are important objects of the chace with the northern *Afiatics*, for their ufes are confiderable. The flefh and fat are efteemed by the natives among the greateft delicacies. Doctor *Pallas* thought the lamb excellent; but the flefh, and efpecially the fat, of the old ones lefs agreeable, when boiled: but if roafted exceedingly good. The fkins, with their winter coat, ferve as warm raiments and coverlets: the horns for variety of neceffaries.

Chace.

The chace of thefe animals is both dangerous and difficult. As foon as they fee a man, they afcend to the higheft peaks of the rocks; and are fhot with the utmoft ftratagem, by winding round the rocks, and coming on them unaware. At other times they are taken in pit-falls made in the paths which lead to their favorite falt or licking-places. Elks, ftags, and roes, and other wild beafts, are taken in thefe pits. They are oft'times fhot with crofs-bows, placed in the way of their haunts, which difcharges its

arrow

arrow whenever the beaft treads on a ftring faftened for that purpofe to the trigger. The *Mongols* and *Tungufi* ufe frequently a nobler method of chace, and furround them with horfes and dogs. The *Kamtfchatkans* pafs the latter part of the fummer to *December*, with all their families, amidft the mountains, in pur-fuit of thefe animals *. The old rams are of vaft ftrength. Ten men can fcarcely hold one. The young are very eafily made tame. The firft trial probably gave origin, among a gentle race of mankind, to the domefticating thefe moft ufeful of quadru-peds: which the rude *Kamtfchatkans* to this moment confider only as objects of the chace, while every other part of the world enjoy their various benefits, reclamed from a ftate of nature.

Befides the notices before cited, taken of thefe animals by the antients, I may add, that *Varro* informs us, that in his days there were wild fheep in *Phrygia* †. *Strabo* fpeaks of the rams of *Sar-dinia*, which have hair inftead of wool, and are called *mufmones* ‡. Of their fkins were made both breaft-plates and cloathing.

The antients did not neglect experiments whether they could not improve the breed. *Columella* ‖ fays, that his uncle, M. *Columella*, a man of ftrong fenfe, and an excellent farmer, pro-cured fome wild rams, which had been brought among other cattle to *Cales* from *Africa*, by way of tribute, which were of a very fingular color. Thefe he turned to his common fheep. The firft produce was lambs with a rough coat, but of the fame color with the rams. Thefe again produced, from the *Tarentine* ewes, lambs with finer fleeces; and in the third generation, the fleeces

* Hift. *Kamtfchatka*.   † De re ruft. lib. ii. c. i.   ‡ Lib. v. p. 344.
‖ De re ruft. lib. vii. c. 2.

H 2   were

were as fine as thofe of the ewes, but the color the fame with that of the father and grandfather. This breed was the fame which the old *Romans* called *umbri*; or fpurious *. But there had been once a notion, that the animal itfelf was no more than an Hybridous produ&ion.

> *Tityrus* ex ovibus oritur, hircoque parente :
> Mufimonem capra ex vervegno femine gignit †.

14. BEARDED.   Tragelaphus feu **Hirco-cervus,** *Caii* Sibirian Goat. *fyn. quad.* No. 11. ed. 1ft. *opufc.* 59.

SH. with the hairs on the lower part of the cheeks and upper jaws extremely long, forming a divided or double beard : with hairs on the fides, and body fhort : on the top of the neck longer, and a little ere&. The whole under part of the neck and fhoulders covered with coarfe hairs, not lefs than fourteen inches long. Beneath the hairs, in every part, was a fhort genuine wool, the rudiments of a fleecy cloathing : the color of the breaft, neck, back, and fide, a pale ferruginous. Tail very fhort.

Horns clofe at their bafe ; recurvated ; twenty-five inches long ; eleven in circumference in the thickeft place ; diverging, and bending outwards ; their points being nineteen inches diftant from each other.

PLACE.   I bought the fkin of this animal in *Holland*. The perfon who fold it, informed me that it came from the *Eaft Indies* : but I

---

* *Plin.* Hift. Nat. lib. viii. c. 49.
† An old epigram quoted by *Hardouin,* on the above paffage in *Pliny.*

8

                                                    rather

Bearded Sheep. — N.° 14.

rather imagine it was brought from *Barbary*, it being probably the fame with the *Lerwee* or *Fifhtal* of Doctor *Shaw* *; who fays, that his *Lerwee* is a moft timorous animal, plunging down the rocks and precipices when purfued.

The fame animal was brought into *England* from *Barbary* in 1561, and well defcribed by my countryman, Doctor *Kay* or *Caius*. He fays, that it inhabited the mountanous and rocky parts of *Mauritania*; and feemed in confinement to be very gentle: full of play, and frolickfome, like a goat. The horns were like thofe of a ram. ~ They were larger in all refpects than thofe I defcribe, fo belonged to a larger-fized animal, which he defcribes to be three feet and a half high to the mane: its whole length, four feet and a half. Under fide of the neck covered with very long hairs, falling as low as the knees: the knees covered alfo tranfverfely with long and thick hair, to preferve them from injury from falls, in any of its vaft leaps. In my fpecimen, thofe parts were guarded by a callus : perhaps the hairs were rubbed off.

The fkin I purchafed was defective about the face. I could not therefore remark nor underftand the *divided beard* defcribed by Doctor *Caius*, till I met with a very fine print, engraven by *Bafan*, from a painting by *Oudry*, taken from the living animal in the *French* King's menagery. From the print it appears that there was no beard on the chin; but that it was formed in the manner I defcribe by the affiftance of the engraving, which fupplied me with the idea given by the learned phyfician.

This I believe to be the *Tragelaphus* of *Pliny* †, not only on account of its beard, and the great length of hair on its fhoul-

* Travels, 243. † Lib. viii. c. 33.

ders;

ders; but likewife of the place where that *Roman* naturalift fays
it was found, near the river *Phafis*; for I am informed by Doctor
Pallas, that an animal with a divided beard, probably the fame,
has lately been difcovered by Profeffor *Guildenftaedt,* on the mountains of *Caucafus*; from whofe foot * arifes the very river, on whofe
banks were its antient haunts.

This fpecies and the laft agree greatly together, the beard
excepted, and great length of hair on the breaft.

* *D'anville.*

Horns

Horns bending backward, and almoſt cloſe at their baſe.   IV. GOAT.
Eight cutting teeth in the lower jaw, none in the upper.
The male bearded.

Ibex. *Plinii lib.* viii. c. 53.   fis, in dorſum reclinatis, gula barbata.   15. IBEX.
Bouc eſtain. *Belon. obs.* 14. Bouc ſau-   *Lin. ſyſt.* 95. *Klein quad.* 16.
vage. *Gaſton de Foix.* 99. Capricorne.   Le Bouquetin. *de Buffon,* xii. 136. *tab.*
Munſter Coſmogr. 381.   xiii. xiv. *Zimmerman.* 114.
Ibex. *Geſner quad.* 303. *Raii ſyn. quad.*   Steinbock. *Kram. Auſtr.* 321. *Ridinger*
77. *Briſſon quad.* 39.   *kleine Thiere,* No. 71. *Br. Muſ. Aſhm.*
Capra Ibex. C. Cornibus ſupra nodo-   *Muſ.* LEV. MUS.

G. with large knotted horns, reclining backwards; ſometimes
three feet long. Eyes large, head ſmall. Male furniſhed with a duſky beard. Hair rough. Color a deep brown,
mixed with ſome hoary. Legs partly black, partly white. Space
under the tail, in ſome tawny, in others white. Belly of a tawny-
white. Tail ſhort. Body ſhort, thick, and ſtrong. Legs ſtrong.
Hoofs very ſhort.

Females are leſſer than the males; have ſmaller horns, like
thoſe of the common ſhe-goat; and have few knobs on the upper
ſurface.

In *Europe,* inhabits the *Carpathian* and *Pyrænean* mountains; and   PLACE.
on the higher piers of the *Sierra de Ronda,* in the province of
Granada* ; in the *Griſons* country; and in the *Vallais,* amidſt the
higheſt points of the *Rhætian Alps,* amidſt ſnow and glacieres. They
are exceſſively wild, and difficult to be ſhot: in very ſevere weather

* *Carter's Hiſt. Malaga.*

3   deſcend

descend a little, in quest of pasturage.   The males, during the time of rutting, bray horribly.   The females, at the time of parturition, separate from the males, and retire to the side of some rill to bring forth: have one, or at most two, at a time.

Their chace very difficult and dangerous: being very strong, they sometimes tumble the incautious huntsmen over the precipices, except they have time to lie down, and let the animals run over them.

It is said, that if they are hard pressed, and cannot escape otherwise, they will fling themselves down the steep precipices, and fall on their horns so as to escape unhurt.   Certain it is, that they are often found with one horn, the other being broken by the fall *.   Some pretend, that to get out of the reach of the huntsmen, they will hang by their horns over the precipices, by a projecting tree, and remain suspended till the danger is past.

Their flesh is esteemed good.   Their blood was once in great repute in pleurisies.   They are said not to be long-lived.

It is found in *Asia*, on the rude summits of that chain of mountains from *Taurus*, continued between eastern *Tartary* and *Sibiria*.   It likewise inhabits the tract beyond the *Lena*; and perhaps *Kamtschatka*: and a few are found to the east of the *Jenesei*. The *Tartars* call them *Tau Tokkè*, or mountain goats.   The horns of these seem more incurvated than those of the *European*; otherwise they agree.

This animal also inhabits the province of *Hedsjæs*, in *Arabia* †, and is called there, *Bæden*.

Lastly, it is found in the high mountains of *Crete*; where

* *Pallas.*   † *Forskal.* iv.

Belon

Belon fays, that if one of them is wounded by an arrow, it cures itfelf by browzing the herb dittany. *Pliny* fays, that ftags extract the fteeled inftrument by the fame remedy *. He fpeaks much of their amazing agility.

The former writer informs us, that there are two fpecies of thefe animals, and that he had feen the horns of each. This is now verified. The fecond I call the *Caucafan*, being lately difcovered by profeffor *Guildenftaedt* on that vaft chain of mountains.

<table>
<tr><td>*Pafen*; Capricerva, *Kæmpfer*, Amæn. exot. 398.</td><td>Act. Petrop. Acad. 1779. p. 273.</td><td>16. CAUCASAN.</td></tr>
<tr><td>Wild goat, *Tavernier's Trav.* ii. 153.</td><td>Æegagrus. Pallas Spicil. Zool. xi. 45. tab. v. fig. 2, 3.</td><td></td></tr>
<tr><td>onardus de Lap. Bezoar 8.</td><td>*Zimmerman,* 662. Mus. Lev.</td><td></td></tr>
</table>

G. with fmooth black horns, fharply ridged on their upper parts, and hollowed on their outward fides. No veftiges of knots or rings, but on the upper furface are fome wavy rifings: bend much back, like thofe of the laft; are much hooked at the end; approach a little at the points. Length three feet. Are clofe at the bafe: one foot diftant in the wideft part: only eight inches and a half from tip to tip. The weight of a pair in the *Leverian* Mufeum weighed ten pounds.

On the chin a great beard, dufky, mixed with chefnut. Forepart of the head black, the fides mixed with brown; the reft of the animal grey, or grey mixed with ruft-color. Along the middle of

BEARD.

* Hift. Nat. lib. viii. c. 27.

the back, from the neck to the tail, is a black lift. The belly, infide of the limbs, and fpace beneath the tail, white. The tail alfo black.

The female is either deftitute of horns, or has very fhort ones, and is beardlefs.

In fize it is fuperior to the largeft he-goats, but in form and agility refembles a ftag: yet *Monardus* compares it to the he-goat, and fays that it has the feet of the goat.

Inhabits the loftieft and moft rude points of *Caucafus*, among the fchiftous rocks, and chiefly about the rivers *Kuban* and *Terek*. All *Afia Minor**, and perhaps the mountains of *India*. They abound on the inhofpitable hills of *Laar* and *Khorazan* in *Perfia*; and accord- ing to *Monardus* are alfo found in *Africa*. They may likewife be found in *Crete*, and even on the *Alps*; for I find among the figures of animals by that great artift *Ridinger*, one † whofe horns bear a refemblance to thofe in queftion. The *Tartars* and *Georgians* make ufe of their horns for drinking cups, and highly efteem their flefh.

It is an animal of vaft agility. *Monardus* was witnefs to the manner of its faving itfelf from injury by falling on its horns; for he faw that which he defcribes leap from a high tower, pre- cipitating itfelf on its horns; then fpringing on its legs, and leaping about, without receiving the left harm. They go to rut in *November*, and bring forth in *April*, therefore, like the common goat, are with young five months.

This is one of the animals which yields the once-valued alexi- pharmic, the Bezoar-ftone; which is a concretion formed of many

---

* *Nov. Com. Petrop.* xx. 452.   † *Entwurf Einiger Thiere,* 71.

coats, incrufting a nucleus of fmall pebble, ftones of fruits, bits of ftraw, or buds of trees. The incrufting coats are created from the vegetable food of the animals, efpecially the rich, dry, and hot herbs of the *Perfian* and *Indian* mountains. Its virtues are now exploded, and it is reckoned only an abforbent, and that of the weakeft kind.

The orientalifts. call the true kind *Pafahr*, from the word *Pafen*, the name of an animal which produces it in *Perfia*; and from *Pafahr* is derived the word Bezoar *. It is produced from numbers of animals; from tame goats, cows, antelopes, deer, *Lama*, *pacos*, and even porcupines, and the apes of *Macaffar* †. Thofe which are procured from the *American* animals are called *occidental*, and were leſt efteemed. But the oriental were fo highly valued, that *Tavernier* fold one, weighing $4\frac{1}{2}$ oz. for 2000 livres.

Since the difcovery of this fpecies of goat, to it muft be given the origin of the tame, as there is the greateft conformity between its horns and thofe of the domeftic kinds; unlefs we can fuppofe that the latter, from their way of life, have loft the knots, the great character of the *ibex*, which I once fuppofed to be their only ftock. I cannot help thinking with Doctor *Pallas*, that they may be derived from both, efpecially as we are affured that an union between the *ibex* and fhe goats will produce a fruitful off-fpring ‡ : yet Mr. *Guldenftaedt* fays that the mountaineers of *Caucafus* never have obferved them to mix or couple with the common goats. I will therefore now proceed to the tame goat, and all its varieties.

THIS ONE STOCK OF THE TAME GOATS.

---

* *Kæmpfer.*   † *Tavernier*, ii. 154.   ‡ *Pallas* Sp. Zool. xi. 48.

a DOMESTIC.

α DOMESTIC. Capra, *Gesner quad.* 266.   Siege Klein quad. 15.
   Raii syn. quad. 77.   Le Bouc, la chevre de *Buffon.* v. 59.
C. hircus, *Lin. syst.* C. cornibus carinatis   *Brisson quad.* 38.
   arcuatis. 94.   Goat, *Br. Zool.* i. Nᵒ 5.
Gef. *Faun. suec.* Nᵒ 44.   Siegen Bock,

The horns of the tame goats have a curvature outwards towards their ends. I have a pair belonging to a *Welsh* he-goat
three feet five inches long, and three feet two inches between tip
and tip. The color of the domestic goats varies : the hair in
some long : in those of hot countries smooth and short.

PLACE.   Inhabits most parts of the world, either native or naturalized :
bears all extremes of weather; being found in *Europe* as high as
Wardhuys in *Norway*, where they breed and run out the whole
year; but in winter only have, during night, the shelter of
hovels : feed in that season on moss and the bark of fir-trees, and
even of the logs cut for fuel. Their skins in *Norway* and
West Bothnia an article of commerce*. Thrive equally well in
the hottest part of *Africa* †, and in *India*, and its islands ‡.

It is not a native of the new world, having been introduced
there first by the discoverers of that continent; for the *Ameri-
cans* were unacquainted with every domestic animal, with sheep,
goats, hogs, cows, and horses‖. The increase of these animals in

* Doctor SOLANDER.
† *Bosman.* 227.
‡ *Dampier,* i. 320.   *Beeckman's* voyage to *Borneo,* 36.
‖ *Ovalle's hist. Chile.   Churchill's coll.* iii. 43.   *Jacques Carthier's voy. Canada.*
Hackluyt's coll. iii. 233.

                                                         all

all parts, especially on the southern tract of that continent, is prodigious; but in the rigorous climate of *Canada* the animal in question is too delicate to perpetuate its race *; so that new supplies are annually imported to prevent its extinction. We mention this, as an agreeable essayist on husbandry †, and the *Swedish* naturalist ‡, have given to *America* animals to which it has no clame.

No animal seems so subject to varieties (the dog excepted) as the goat; *Capræ tamen in multis similitudines transfigurantur*, is a very just observation of *Pliny* §; for besides those of *Britain* and *France*, are the following, that differ extremely from each other: at the head of these should be placed one not less eminent for its beauty than its use.

VARIETIES.

β ANGORA: *Lin. syst.* 94. *De Buffon*, v. 71. *Brisson quad.* 39. *Zimmerman,* 134. LEV. MUS.

A variety that is confined to very narrow bounds; inhabiting only the tract that surrounds *Angora* and *Beibazar,* towns in *Asiatic Turkey* ‖, for the distance of three or four days journey. *Strabo* ╪ seems to have been acquainted with this kind; for speaking of the river *Halys*, he says, that there are goats found near it that are not known in other parts.

* *De Buffon*, ix. 71.
† P. 137.
‡ *Syst. nat.* p. 95. sp. 6. & 7.
§ *Lib.* viii. c. 53.   ‖ *Tournefort's voy.* ii. 351.   ╪ *Lib.* xii. p. 823.

In

GOAT.

In the form of their body they differ from the common goat, being shorter; their legs too are shorter, their sides broader and flatter, and their horns straiter; but the most valuable characteristic is their hair, which is soft as silk, of a glossy silvery whiteness, and curled in locks of eight or nine inches in length.

This hair is the basis of our fine camlets, and imported to *England* in form of thread; for the *Turks* will not permit it to be exported raw, for a reason that does them honor; because it supports a multitude of poor, who live by spinning it *.

The goatherds of *Angora* and *Beibazar* are extremely careful of their flocks, frequently combing and washing them. It is observed, that if they change their climate and pasture, they lose their beauty; we therefore suspect that the design of Baron *Alstroemer*, a patriotic *Swede*, turned out fruitless, who imported some into his own country, to propagate the breed, for the sake of their hair.

We imagine that the goats of *Cougna* (the old *Iconium*) are varieties of the *Angora* kind; for *Tournefort* mentions them together, and says the former are preferred because the latter are all either brown or black.

The horns of the he-goat do not bend, but stand diverging from each other; their length is two feet one; the space between tip and tip two feet ten and a half; they are twisted spirally, in a most elegant manner. The horns of the female bend back, and are short.

* *Haffelquift's voy. Eng. transl.* 191. *Tournefort voy.* ii. 351. According to *Nieuhoff* they are also found at *Gomron, Churchill's coll.* 232.

γ SYRIAN,

X

Syrian Goats.

γ SYRIAN. Capra mambrina feu fyriaca. *Briffon quad.* 47.
Gefner quad. 153. *Raii fyn. quad.* 81. *Profper Alp. hift. Ægypti,* i. 229.
C. cornibus reclinatis, auribus pen- *Rauwolff's travels,* ii. 71. *Ruffel's Aleppo,*
dulis, gula barbata. *Lin. fyft.* 95. 62. *Zimmerman,* 135.

Plentiful in the *Eaft:* fupply *Aleppo* with milk. Their ears of a vaft length, hanging down like thofe of hounds : are from one to two feet long : fometimes they are fo troublefome, that the owners cut off one to enable the animal to feed with more eafe. The horns are black and fhort.

The fame fpecies is alfo found among the *Kirghifian Tartars,* and fometimes brought down to *Aftracan.*

δ AFRICAN. Capra depreffa. C. cornibus Le bouc d'Afrique. *De Buffon,* xiii. 154.
erectis apice recurvis. *Lin. fyft.* 95. *tab.* xviii. xix. LEV. MUS.

A dwarf variety, found in *Africa.* The male covered with rough hair, and beneath the chin hang two long hairy wattles: the horns fhort, very thick, and triangular, and lie fo clofe to the fcull as almoft to penetrate it : the horns of the female are much lefs, neither has it wattles : its hair is fmooth.

ε WHIDAW. Capra reverfa. C. cor- *fyft.* 95.
nibus depreffis incurvis minimis cra- Le bouc de Juda. *De Buffon,* xii. 154.
nio incumbentibus, gula barbata. *Lin.* *tab.* xx. xxi.

From *Juda* or *Whidaw,* in *Africa.* A fmall kind: the horns fhort, fmooth, and turn a little forwards. *Linnæus* fays, that this
and

and the preceding came from *America*; but certainly, before its difcovery by the *Spaniards*, the goat and every other domeftic animal was unknown there.

♌ Capricorn.   Le Capricorne.   *De Buffon*, xii. 146. *tab.* xv.

A variety with fhort horns, the ends turning forward : their fides annulated : the rings more prominent before than behind.

In the country of the *Cabonas*, north of the Cape of *Good Hope*, is a fpecies of tame goats refembling the common kind, only that they want horns*.

17. Pudu.   Le Pudu *Molina Chili* 291.   *Ovis Pudu Gmelin Lin.* 201.

G. with brown hair; round fmooth horns turning outwards: fize of a kid fix months old : no beard; in all other refpects has quite the characters of the goat.

Place.   Inhabits the *Andes*; defcends at approach of winter, in vaft herds, to feed on the fouthern plains of *Chili*. The Chilians catch them in great numbers, not only for food, but for the fake of rearing them, in which they have great fuccefs : they are gentle animals, and very foon domefticated.

* Journal hiftorique, 76.

Horns

XI

Giraffa, or Camelopard.— S.18.

Horns short, upright, truncated at the top.

Neck and shoulders of a vast length.

Eight cutting teeth in the lower jaw, the two outmost bilobated. No teeth in the upper jaw.

Camelopardalis. *Plinii lib.* viii. *c.* 18. *Dion Cassius, lib.* xliii. *Præneft. pavem. apud Shaw suppl.* 88. *Oppian cyneg.* iii. 466. La Giraffe que les *Arabes* nomment Zurnapa. *Belon obs.* 118. 119. *Leo Afr.* 337. *Gesner quad.* 160. *Raii syn. quad.* 90. *Brisson quad.* 37. *De Buffon,* xiii. 1.

Cervus camelopardalis. C. cornibus simplicibus, pedibus anticis longissimis. *Lin. syst.* 92. Tragus Giraffa. *Klein. quad.* 22. *Zimmerman,* 534. *Sparman's* voy. ii. 149. 237. *Paterson's Travels.* 125.

G. with short strait horns covered with hair, and truncated at the end and tufted: in the forehead a tubercle, about two inches high, resembling a third horn. The length, according to the measurement given by Mr. *Hop* in his *journal historique,* p. 28, from the nose to the tip of the tail above eighteen feet. Height from the crown of the head to the soles of the fore feet seventeen feet: from the top of the rump to the bottom of the hind feet only nine: length of the neck seven: from the withers to the loins only six: the fore legs not longer than the hind legs; but the shoulders of a vast length, which gives the disproportionate height between the fore and hind parts: the chest extremely projecting, and almost tuberous: head resembles that of a stag: the neck slender and elegant: on the upper part is a short erect mane: the ears large: horns, according to Mr. *Paterson,* one foot and half an inch long, ending abrupt, and with a tuft of hair issuing from the summit: they are not deciduous.

VOL. I.   K   The

The height of that killed by Mr. *Paterſon* was only fifteen feet. The head is of an uniform reddiſh brown: the neck, back, and ſides, outſides of the ſhoulders and thighs varied with large teſſellated, dull ruſt-colored marks of a ſquare form, with white ſeptaria, or narrow diviſions: on the ſides the marks are leſs regular: the belly and legs whitiſh, faintly ſpotted: the part of the tail next to the body is covered with ſhort ſmooth hairs, and the trunk is very ſlender: towards the end the hairs are very long, black, and coarſe; and forming a great tuft hanging far beyond the tip of the trunk: the hoofs are cloven, and nine inches broad, and black. This animal wants the ſpurious hoofs.

The female has four teats. Mr. *Paterſon* ſaw ſix of theſe animals together; poſſibly they might have been the male and female, with their four young.

PLACE, AND MANNERS.

Inhabits the foreſts of *Æthiopia*, and other interior parts of *Africa*, almoſt as high as *Senegal*; but is not found in *Guinea*, or any of the weſtern parts; and I believe not farther ſouth than about *lat.* 28. 10*, among the *Nemaques* on the northern ſide of the *Orange* river. It is very timid, but not ſwift: from the ſtrange length of its fore legs, cannot graze without dividing them to a vaſt diſtance; it therefore lives by brouzing the leaves of trees, eſpecially that of the *mimoſæ* and a tree called the wild apricot: kneels like a camel when it would lie down; and is a gentle animal. When it would leap, it lifts up its fore legs and then its hind, like a horſe whoſe fore legs are tied. It runs very badly and aukwardly, but continues its courſe very long before it ſtops. It is very difficult to diſtinguiſh this animal at a diſtance, for when ſtanding they look like a decayed tree by reaſon of their form, ſo are

* Journal hiſtorique, &c. 24.

vaſſed

paſſed by, and by that deception eſcape. I ſaw the ſkin of a young one at *Leyden*, well ſtuffed, and preſerved; otherwiſe might poſſibly have entertained doubts in reſpect to the exiſtence of ſo extraordinary a quadruped. *Belon*'s figure very good.

Known to the *Romans* in early times; appears among the figures in the aſſemblage of *eaſtern* animals on the celebrated *Prœneſtine* Pavement, made by the direction of *Sylla*, and is repreſented both grazing and brouzing, in its natural attitudes : was exhibited at *Rome* by the popular *Cœſar*, among other animals in the *Circœan* games. Finely and juſtly deſcribed by *Oppian*.

K 2   Annulated

VI. ANTELOPE.   Annulated or twisted horns.

Eight broad cutting teeth in the lower jaw; none in the upper.

Inside of the ears marked lengthways with three feathered lines of hair.

Limbs of a light and elegant form.

THE several species that compose this genus, two or three excepted, inhabit the hottest part of the globe; or at least those parts of the temperate zone that lie so near the tropics as to form a doubtful climate.

None therefore, except the *Saiga**, and the *Chamois*, are to be met with in *Europe*; and, notwithstanding the warmth of *South America* is suited to their nature, yet not a single species has ever been discovered in any part of the new world. Their proper climates seem therefore to be those of *Asia* and *Africa*, where the species are very numerous.

As there appears a general agreement in the nature of the species that form this great genus, it will prevent a needless repetition, to observe here, that the ANTELOPES are animals generally of a most elegant and active make; of a restless and timid disposition; extremely watchful; of great vivacity; remarkably swift, re-

* Found between the *Don* and *Dnieper*; and, as I have heard, even *Transylvania.*

markably

markably agile; and moſt of their boundings ſo light, ſo elaſtic, as to ſtrike the ſpectator with aſtoniſhment. What is very ſingular, they will ſtop in the midſt of their courſe, for a moment gaze at their purſuers, and then reſume their flight*.

As the chace of theſe animals is a favorite diverſion with the *eaſtern* nations, from that may be collected proofs of the rapid ſpeed of the ANTELOPE tribe. The Grehound, the fleeteſt of dogs, is uſually unequal in the courſe; and the ſportſman is obliged to call in the aid of the Falcon, trained to the work, to ſeize on the animal and impede its motions, to give the dogs opportunity of overtaking it. In *India,* and in *Perſia,* a ſort of Leopard is made uſe of in the chace: this is an animal that takes its prey not by ſwiftneſs of foot, but by the greatneſs of its ſprings, by motions ſimilar to that of the ANTELOPE; but ſhould the Leopard fail in its firſt eſſay, the game eſcapes†.

The fleetneſs of this animal was proverbial in the country it inhabited even in the earlieſt times: the ſpeed of *Aſahel* ‡ is beautifully compared to that of the ‖ *Tzebi*; and the *Gadites* were ſaid to be as ſwift as the *Roes* upon the mountains. The ſacred writers took their ſimilies from ſuch objects as were before the eyes of the people they addreſſed themſelves to. There is another inſtance drawn from the ſame ſubject; the diſciple raiſed to life at *Joppa* was ſuppoſed to have been called *Tabitha,* i. e. *Dorcas,* or the ANTELOPE, from the beauty of her eyes; and

* *Shaw's trav.* 244.
† *Bernier's trav.* iv. 45. *Voy. de Boullaye le Gouz,* 248.
‡ 2 *Sam.* ii. 18.
‖ *Shaw's trav. ſuppl.* 74; who informs us, that this word ſhould have been tranſlated, the *Antelope*; not the *Roe,* as the text has it.

this

this is ſtill a common compariſon in the *Eaſt: Aine el Czazel*, or " You have eyes of an ANTELOPE," is the greateſt compliment that can be paid to a fine woman *.

Some ſpecies of the ANTELOPES form herds of two or three thouſands, while others keep in ſmall troops of five or ſix. They generally reſide in hilly countries; though ſome inhabit plains: they often brouze like the goat, and feed on the tender ſhoots of trees, which gives their fleſh an exellent flavor. This is to be underſtood of thoſe that are taken in the chace; for thoſe that are fattened in houſes are far leſs delicious. The fleſh of ſome ſpecies are ſaid to taſte of muſk, which perhaps depends on the qualities of the plants they feed on.

This preface was thought neceſſary, to point out the difference in nature between this and the Goat kind, with which moſt of the ſyſtematic writers have claſſed this animal: but the ANTELOPE forms an intermediate genus, a link between the Goat and the Deer. They agree with the firſt, in the texture of the horns, which have a core in them; and they never caſt them: with the laſt, in the elegance of their form, and great ſwiftneſs.

* with hooked horns.

19. GNOU.   Bos Gnou. *Zimmerman*, 372. *Journal Hiſt.* 53. tab.   p. 54. LEV. MUS.

HORNS.   A. with horns ſcabrous, and thick at the baſe, bending forward cloſe to the head, then ſuddenly reverting upwards: the ends ſmooth. baſes two inches diſtant: tips one foot three: length along the curve one foot five. The females are horned

* *Pr. Alp. hiſt. Ægypt.* i. 232.

exactly

exactly like the males *. Horns in the young animals quite ftrait.

Mouth fquare; upper and lower tip covered with fhort ftiff **HEAD.**
hairs: the lower with long briftles intermixed. Noftrils covered
with broad flaps. From the nofe, half way up the front, is a
thick oblong-fquare brufh of long ftiff black hairs reflected upwards, on each of which the other hairs are long, and point
clofely down the cheeks. Round the eyes are difpofed in a radiated form feveral ftrong hairs.

Neck fhort, and a little arched. On the top a ftrong and up- **NECK.**
right mane, reaching from the horns beyond the fhoulders. On
the chin a long white beard; and on the gullet a very long pendulous bunch of hair. On the breaft, and between the fore legs,
the hairs are very long, and black.

Tail reaches to the firft joint of the legs, and is full of hair **TAIL.**
like that of a horfe, and quite white.

The body is thick; and covered with fmooth fhort hair of a **COLOR.**
rufty brown color tipt with white.

Legs long, elegant, and flender, like thofe of a ftag. On each
foot is only a fingle fpurious or hind hoof.

The height of one brought over to the *Hague* was three feet **SIZE.**
and a half. The length from between the ears to the *anus* fix
and a half: but they grow to a greater fize.

It is a ftrange compound of animals: having a vaft head like
that of an ox: body and tail like a horfe: legs like a ftag: and
the *finus lacrymales* of an antelope.

The flefh is of a very fine grain, very juicy and of a moft delicate flavor, in tafte refembling that of others of the genus, and
without the left refemblance to that of beef.

* *Sparman.*

4 It

It inhabits in great herds the fine plains of the great *Namacquas*, far north of the Cape of *Good Hope*, extending from S. lat. 25. to 28. 42. where *Africa* feems at once to open its vaft treafures of hoofed quadrupeds. It probably may be found higher, but as yet that is uncertain.

It is exceedingly fierce, and ufually on the fight of any body drops its head and puts itfelf into an attitude of offence: and will dart with its horns againft the pales of the inclofure towards the perfons on the outfide; yet will afterwards take the bread which is offered. It will often go upon its knees, run fwiftly in that fingular pofture, and furrow the ground with its horns and legs.

The *Hottentots* call it *Gnou* from its voice. It has two notes, one refembling the bellowing of an ox, the other more clear. It is called an ox by the *Europeans*. I therefore fufpect the wild grey ox, of great fwiftnefs, defcribed by *Leo*, to be of this kind; and perhaps the *Baas*, p. 36 of this work.

---

20. CHAMOIS.

Rupicapra, *Plinii* lib. viii. c. 15. *Gefner quad.* 290. *Raii fyn. quad.* 78. *Scheuchzer. It. Alp.* i. 155. &c.
Capra rupicapra. C. Cornibus erectis uncinatis. *Lin. fyft.* 95.
Chamois ou Yfard. *Belon obf.* 54.
Yfarus ou Sarris. *Gafton de Foix,* 99 *.

Briffon quad. 41. *de Buffon,* xii. 136. *tab.* xvi.
Gemfe, *Klein quad.* 18. *Ridinger Kleine Thiere,* No. 72. *wild Thiere,* 25.
Antilope rupicapra. *Pallas mifcel.* 4. *Spicil.* xii. 12. LEV. MUS.

G. with flender, black, upright horns, hooked at the end: behind each a large orifice in the fkin: forehead brown: cheeks, chin, and throat white: belly yollowifh: reft of the body

---

* *Gafton de Foix, Seigneur du Rù,* commonly called *Roy Phebus,* a celebrated writer on hunting, whofe works are added to thofe of *Jaques de Fouilloux,* entitled, *La Venerie & Fauconnerie. Paris.* 1585.

deep

deep brown: hair long: tail short: hoofs much divided, short and goat-like.

In some (differing perhaps in sex) the cheeks and chin are dusky, and the forehead white.

Inhabits the *Alps* of *Dauphinè*, *Switzerland*, and *Italy*; the *Pyrænean* mountains, the *Sierra de Ronda*, *Greece*, *Crete*, and the mountains of *Caucasus* and *Taurus*.   It does not dwell so high in the hills as the *Ibex*, and is found in greater numbers.   They feed before sunrise and after sun-set: during winter lodge in hollows of the rocks, to avoid the falls of the *Avelenches* : during that season, eat the slender twigs of trees, or the roots of plants, or herbs, which they find beneath the snow: are very timid and watchful: each herd has its leader, who keeps centry on some high place while the rest are at food; and if it sees an enemy, gives a short sort of a hiss by way of signal, when they instantly take to flight.

They have a most piercing eye, and quick ear and scent: are excessively swift and active: are hunted during winter for their skins, which are very useful in manufactures, and for the flesh, which is very well tasted.   The chace is a laborious employ: they must be got at by surprize, and are shot with riflebarrel'd guns.   In their stomachs is often a hairy ball, covered with a hard crust of an oblong form: are said to be long lived: bring two, seldom three, young at a time.

PLACE.

CHACE.

** With arcuated horns.

Blue Goat. *Kolben's Cape* ii. 114.   *Spicil. Zool.* 6. *Br. Muſ.* Lev. Mus.
Antelope Leucophœa. *Pallas Miſcel.* 4.   Le *zeiran de Buffon ſuppl.* vi. 168.

A. with ſharp-pointed, taper, arcuated horns, bending backwards, marked with twenty prominent rings, but ſmooth towards their points; twenty inches long: ears ſharp-pointed, above nine inches in length. Larger than a buck. Color, when alive, a fine blue, of a velvet appearance: when dead, changes to a blueiſh-grey, with a mixture of white. The hairs long. Beneath each eye is a large white mark. The belly white. The tail ſeven inches long; the hairs at the end ſix inches.

In ſize, ſuperior to the fallow deer or buck.

I deſcribed it from a ſkin which I bought at *Amſterdam*, brought from the *Cape* of *Good Hope*. I was informed, that they are found far up the country, north of that vaſt promontory; which I find confirmed by the late journies*. It is called by the *Dutch* the *Blauwe Bock*, or blue goat.

M. *de Buffon* deſcribes it under the ſame name, *ſuppl.* vi. 194. and in p. 168. again under the improper *Aſiatic* name of *Tzeiran*, which belongs to a very different ſpecies, the *Chineſe*, No. 36. but has borrowed the figure from the *Dutch* travellers.

This is the ſpecies, which, from the form of the horns and length of the hair, ſeems to connect the Goat and Antelope race.

* Journal Hiſtorique, &c. *Amſterdam*, 1778, p. 58, where it is called *Bouc-chamois*; and a good figure given of it.

*** Strait

*** Strait horns.

Gazella indica cornibus rectis longiſſimis nigris prope caput tantum annulatis. *Raii ſyn. quad.* 79.

Capra Gazella. C. cornibus teretibus rectiſſimis longiſſimis annulatis. *Lin. ſyſt.* 96.

Antelope Bezoartica. *Pallas, ſp. Zool.* i.

14. Ant. oryx. xii. 16.

Le Paſan. *journal hiſtorique,* 56.

La Gazelle des Indes. *Briſſon quad.* 43.

Le Paſan. *De Buffon,* xii. 213. tab. xxxiii. fig. 3. xv. 190. *Br. Muſ. Aſhm. Muſ.* Lev. Mus.

22. ÆGYPTIAN

A. with ſtrait ſlender horns, near three feet long, annulated above half of their length: the reſt ſmooth. Space between horn and horn at the points fourteen inches. At their baſe is a black ſpot; in the middle of the face another; a third falls from each eye to the throat, united to that in the face by a lateral band of the ſame color: the noſe and reſt of the face white. From the hind-part of the head, along the neck and top of the back, runs a narrow duſky line of hairs, longer than the reſt, and ſtanding above them, dilating towards the rump. Sides of a light reddiſh aſh-color; the lower part bounded by a broad longitudinal duſky band, reaching to the breaſt.

Belly, rump, and legs white; each leg marked below the knees with a duſky mark. Tail covered with long black hairs; from the rump to the end of the hairs, two feet ſix inches long.

The length of the ſkin, which I examined, was above ſix feet ſix inches.

Inhabits *Ægypt, Arabia, India,* and the North-weſtern parts of the Cape of *Good-Hope.*

It is ſaid to be a moſt dangerous animal when wounded, nor will the *Hottentots* approach it, unleſs they are ſatisfied that it be totally deprived of life.

23. LEUCORYX.	Antelope Leucoryx cornibus fubulatis rectis, convexe annulatis, corpore lac-	teo? *Pallas fp. Zool.* xii. 16. Oryx, *Oppian. Cyneg.* ii. v. 445.

A. with the nofe thick and broad, like that of a cow. Ears fomewhat flouching. Body clumfy and thick. Limbs lefs fo. Horns long, very flightly incurvated, flender, annulated part of the way: black, pointed. Tail reaching to the firft joint of the legs, and tufted. Color in all parts a fnowy whitenefs, except the middle of the face, fides of the cheeks, and limbs, which are tinged with red.

SIZE.

Size of a *Welfh* runt.

PLACE.

This fpecies inhabits *Gow Bahrein*, an ifle in the gulph of *Baffora*. I difcovered two drawings of the animal in the *British Mufeum*, taken from life in 1712, by order of Sir *John Lock*, agent to the *Eaft India* company at *Ifpahan*. They were preferved as rarities by *Shah Sultahn Houffein*, emperor of *Perfia*, in his baague of *Caffar*, a park eight leagues from the capital *.

A horn, fufpected by Dr. *Pallas* to have belonged to a beaft of this kind, was found foffil in *Sibiria* †.

This animal is probably the *Leucoryx* of *Oppian*, and differs only in wanting the black marks about the temples and cheeks, as mentioned in the following excellent defcription of the poet's, and which Sir *John Lock*'s painter might omit.

---

* The account is taken from a paper attending the drawing.
† *Nov. Com. Petrop.* xiii. 468. tab. x. fig. 5.

4

En

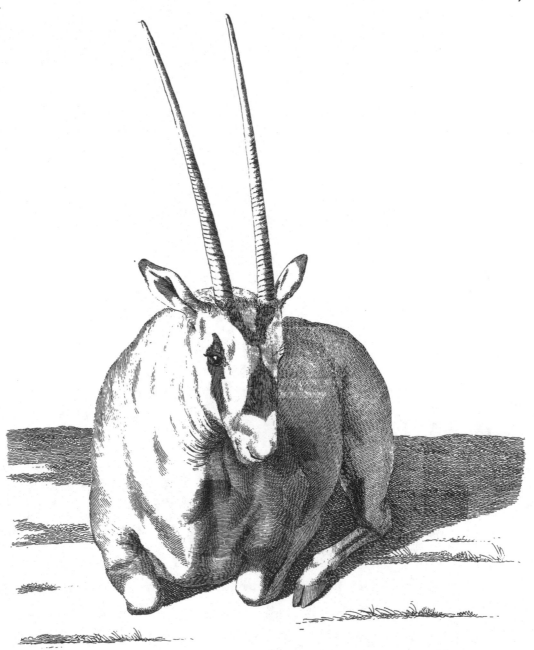

XII

Leucoryx Antelope. — N°23.

En enim fera quæ fylvas perluftrat opacas ;
Cornua acuta ferens animifque ferocibus iram
Formidandus oryx, homines ferafque laceffans ;
Huic candore cutis niveo diftincta relucet
In morem verni lactis ; fed tempora circum
Atque genas nigricat, duplicem pinguedine fpinam
Latè diffindit ; mucrones cornibus atri.

Oppian de Ven. ii. *interpret. Gabr. Bodeno.*

Gornu ignotum. *Gefner quad.* 309.	La Gazelle du Bezoar. *Briffon quad.* 44.	24. ALGAZEL.
La Gazelle. *Belon. obf.* 120. *Alpin. hift.*	Algazel. *De Buffon,* xii. 211. *tab.* xxxiii.	
Ægypt. i. 232. *tab.* xiv.	*fig.* 1. 2.	
Animal bezoarticum. *Raii fyn. quad.*	Capra bezoartica. C. cornibus arcua-	
80.	tis totis annulatis, gula barbata. *Lin.*	
Antelope Gazella. *Pallas, fp. Zool. fafc.*	*fyft.* 96. *Br. Muf. Afhm. Muf.* LEV.	
xii. 16.	MUS.	

A. with very long, flender, upright, horns, bending at the upper part inward towards each other; fome are much annulated, others fmoother. The color red; breaft and buttocks white.

Inhabits *Bengal, Lybia, Ægypt,* and *Æthiopia.* It runs fwiftly up hill, and but flowly along a plain: is very eafily made tame.

Both *Belon* and *Alpinus* note the form of the horns, which they call lunated, or in form of a crefcent.

I never faw any more of this animal than its horns, which are not unfrequent in the cabinets of the curious. They are fuffi-cient to determine me to pronounce the fpecies to be diftinct from the foregoing. *Belon* and *Profper Alpinus* agree in the color, which they declare to be red, and omit all mention of the ftriking, and very characteriftic marks of the other.

25. INDIAN.   Le Coudous. *De Buffon*, xii. 357. *tab.*   An. oreas. *fpic.* xii. 17.
               47.   Pacaffe. *Voy. Congo. Churchill's Coll.* i.
               Antilope oryx. *Pallas fpicil.* 15.   623. *Br. Muf. Afhm. Muf.* LEV. MUS.

A. with thick ftrait horns, marked with two prominent fpiral ribs near two-thirds of their length; fmooth towards their end: fome are above two feet long: thofe at the *Britifh Mufeum*, with part of the fkin adhering, are black. Head of a reddifh color, bounded on the cheeks by a dufky line. Ears of a middling fize. Forehead broad: nofe pointed. On the forehead, a ftripe of long loofe hairs, and on the lower part of the dewlap, a large tuft of black hair.

Along the neck and back, from head to tail, is a black fhort mane: the reft of the body of a blueifh grey, tinged with red. Space between the hoofs and falfe hoofs black.

The tail does not reach to the firft joint of the leg; is covered with fhort cinereous hair; the end tufted with long black hairs.

The hoofs are fhort, furrounded at their junction with the legs, with a circle of black hairs.

SIZE.   The height to the fhoulders is five feet: is thick bodied, and ftrongly made: but the legs are flender.

FEMALES.   The females are horned like the males. This fpecies wants the *finus lacrymalis* *.

NAME.   The *Caffres* call this fpecies *Empofos*. If this is the *Pacaffe*, as there is reafon to fuppofe it to be, they vary in color; the *Pacaffe* being white, fpotted with red and grey. The *Dutch* of the *Cape*

* *Sparman.*

4   call

call it the *Eland* or *Elk*. The *Hottentots*, *t'gann*, from which is formed the name *Canna*. M. *de Buffon*, by mistake, calls this the *Coudous*, which he ought to have bestowed on his *Condoma*.

Inhabits *India*, *Congo*, and the southern parts of *Africa*. Frequents the plains and vallies of the country. They feed chiefly by browsing on shrubs and bushes; and when taken young are soon domesticated. As it is an animal of great strength, it seems possible to render it as useful as the horse or ox, which would be of no small service to the *African* colonists in the neighborhood of the *Cape*, as it is said to be content with a very little food. These animals are in seasons of great drought supposed to migrate from the interior parts of *Africa* in greater numbers than usual. They live in numerous heards; but the old males are often solitary. They grow very fat, especially about the breast and heart: so that they are easily caught: and when pursued, will sometimes fall dead in the chace. Are slow-runners: when roused, always go against the wind, nor can the hunters (even if they front the herd) divert them from their course. The flesh is fine grained, very delicious, and juicy. The hide is tough and thick, especially that of the neck of the male; and is reckoned the best next to that of the Cape buffalo, p. 35, for making of traces, harnesses, or field shoes. The *Hottentots* make tobacco-pipes of the horns.

Ourebi, *Allamand Supplem.* V. 33. tab. xii.   26. OUREBI.

A. with small strait horns, small head, long neck, long pointed ears. Color above, a deep tawny, brightening towards the sides, neck, head and legs; lower part of the breast, belly,

buttocks,

buttocks, and infide of the thighs, white. Tail only three inches long, and black. Hair on the body fhort, under the cheft long and whitifh. On each knee is a tuft of hair. The females are hornlefs.

SIZE.   Length three feet nine to the tail.

PLACE.   Inhabits the country very remote from the *Cape*. Seldom more than two are feen together: they ufually haunt the neigh-borhood of fountains furrounded with reeds. Are excellent ve-nifon.

27. KLIP SPRINGER.   A. oreotragus. *Schreber. tab.* cclix. Gmelin, Lin. 189.

A. with horns quite ftrait, flender, fharp pointed, wrinkled at the bafe, five inches long. Female head hornlefs, round, of a yellowifh grey marked with black rays. Color of the body a yellowifh tawny. Tail very fhort, lies clofe to the body, covered with very fhort hairs, and is fcarcely vifible. Size of a roebuck.

PLACE.   Inhabits the fummits of the higheft and moft tremendous rocks near the *Cape*, and on the fight of man retires to the moft inaccef-fible precipices: and will jump from one crag to another over the moft frightful abyffes. Nothing equals their activity: are fhot with a ball, and are much valued for the fine flavour of the flefh. We are indebted to Doctor *Forfter* for an accurate figure and defcription of this fpecies.

Le Guib. *De Buffon*, xii. 305. 327. *tab.*   *Spicil.* 15. *Sparman*, ii. 219.   28. HARNESSED.
   xl. Antelope fcripta. *Pallas Mifcel.* 8.   Spotted goat, *Kolber*, ii. 115.

A with ftrait horns nine inches long, pointing backwards,
. with two fpiral ribs: ears broad: color a deep tawny:
beneath each eye a white fpot: fides moft fingularly marked with
two tranfverfe bands of white, croffed by two others from the
back to the belly: the rump with three white lines pointing
downwards on each fide: the thighs fpotted with white: tail ten
inches long, covered with long rough hairs.

   Inhabits the plains and woods of *Senegal*, living in large herds.
This is called at the *Cape*, the *Bonte Bock*, or *fpotted goat*. But is
not found farther to the eaft of that part of *Africa* than *Zwel-
lendam*.

Capra fylveftris *Africana* Grimmii. *Raii*   Le Chevrotain d'Afrique. *Briffon quad.*   29. GUINEA.
   fyn. quad. 80. *Klein quad.* 19.   67. *Seb. Muf.* i. *tab.* 43. C. D.
Mofchus Grimmia. M. capite fafciculo   Antilope Grimmia.   *Pallas Mifcel.* 10.
   tophofo. *Lin. fyft.* 92.   .   *tab.* i. *Spicil.* 38. *tab.* iii. LEV. MUS.
La Grimme. *De Buffon*, xii. 307. *tab.* xli.

A with ftrait black horns, flender, and fharp-pointed, not
. three inches long, flightly annulated at the bafe: height
about 18 inches: moft elegant form: ears large: eyes dufky;
below them a large cavity, into which exuded a ftrong-fcented
oily liquid: between the horns a tuft of black hairs. The color
of the neck and body brown, mixed with a cinereous, and a tinge
of yellow: belly white: tail fhort; white beneath, black above.

   I examined this animal a few years ago, in company with
VOL. I.   M   Doctor

Doctor *Pallas*, at the *Prince of Orange*'s menagery, near the *Hague*. Several had been brought over from *Guinea*; but, except this, all died. Dr. *Pallas* said that the females were hornlefs, but are tufted in the fame manner as the males: it feems, therefore, that Dr. *Grimm*, who firft defcribed this fpecies, never faw any but the female.

A beautiful fpecimen of a male, in the LEVERIAN *Mufeum*, is of a bright-bay color. The legs cinereous.

This fpecies extends from *Guinea* to the Cape of *Good Hope*, and is known there by the name of the *Duyker bock*, or *Diving Goat*. It lives always among the brufh wood; and, when it perceives the approach of a man, leaps up, and as fuddenly fquats down; then takes to flight, and every now and then fprings into fight to difcover whether it is purfued.

---

30. ROYAL.

King of the harts, *Bofman's voy.* 236.   *Adanfon's voy.* 207.
Petite biche. *Des Marchais,* i. 312.   Le Chevrotain de Guineé. *De Buffon,*
Cervula parvula Africana. *Seb. Muf.* i.   xii. 315. *tab.* xliii. *fig.* 2. its horn.
70. *tab.* xliii.

A. with very fhort ftrait horns, black and fhining as jet; fcarce two inches long: ears broad: height not above nine inches: legs not thicker than a goofe-quill: color a reddifh brown. The females want horns.

PLACE.

Inhabits *Senegal*, and the hotteft parts of *Africa:* called in *Guinea*, *Guevei*: is very agile, will bound over a wall twelve feet high: is very tame, but fo tender as not to endure tranfportation into our climate.

⁎ Horns

XIII

White-footed Antelope. — N.º 32.

*** Horns bending forwards.

Quadruped from *Bengal*. *Ph. Tr.* No. 775. *Schreber.* cclxii.   31. INDOSTAN.
476. *Alridg.* xi. 898. *tab.* vi.   Antilope Tragocamelus. *Pallas Miscel.* 5.
Biggel. *Mandelslo's voy. Harris's coll.* i.   *Spicil.* 9.

A with horns feven inches long, bending forward: eyes
. black and lively: neck ftrong, bending forward like
that of a camel; along the top a fhort mane: on the fhoulders a
large lump, refembling that of the *Indian* ox, tufted with hair:
hind parts like thofe of an afs: tail 22 inches long, terminated
with long hairs: legs flender: on the lower part of the breaft
the fkin hangs like that of a cow: hair fhort and fmooth, of a
light afh-color, in fome parts dufky; beneath the breaft, and
under the tail, white: on the forehead is a black rhomboidal
fpot. The height of this animal, to the top of the lump on its
fhoulders, was 12 hands.

 Inhabits the moft diftant parts of the *Mogul*'s dominions;   PLACE.
chews the cud; lies down and rifes like a camel: its voice a fort
of croaking, or like the rattle of deer in rutting-time. Doctor
Parfons, to whom we were of late years obliged for the beft
zoologic papers in the *Philofophical Tranfactions*, was the only
writer who has defcribed this animal.

Antelope picta, *Pallas* fpicil. xii. 14. Nyl-ghau. *Ph. Tranf.* lxi. 170. tab. v.   32. WHITE-FOOT-
 Schreber cclxiii. Mus. Lev.   ED.

A with fhort horns, bending a little forward: ears large,
. marked with two black ftripes: a fmall black mane on
the neck, and half way down the back: a tuft of long black hairs

ANTELOPE.

on the fore-part of the neck; above that a large fpot of white; another between the fore-legs on the cheft: one white fpot on each fore-foot; two on each hind-foot: tail long, tufted with black hairs: color a dark-grey.

Female of a pale brown color: no horns: with a mane, tuft, and ftriped ears, like the male: on each foot three tranfverfe bands of black and two of white.

Height to the top of the fhoulders four feet and an inch. Length from the bottom of the neck to the anus four feet*.

Horns feven inches long: triangular towards their bottom; blunt at top. Diftant at their bafes three inches and a quarter; in which they vary from thofe of the Antelope race. Diftant at the points fix inches and a quarter. The head is like that of a ftag.- The legs delicate.

Inhabits the diftant and interior parts of *India,* remote from our fettlements. They are brought down as curiofities to the *Europeans,* and have of late years been frequently imported into *England.* I am not acquainted with the particular part of the country which they inhabit at prefent. In the days of *Aurenge Zebe,* they abounded between *Delli* and *Lahor,* on the way to *Cachemire.* They were called *Nyl-ghau,* or *blue* or *grey bulls:* and

were one of the objects of chace, with that mighty prince, during his journey: they were inclofed by his army of hunters within nets, which being drawn clofer and clofer, at length formed a fmall precinct; into this the king, his *omrahs,* and hunters entered, and killed the beafts with arrows, fpears, or mufquets; and

* Thefe meafurements are taken from the accurate defcription with which Doctor *Hunter* has favoured the public, in the *Philofophical Tranfactions.*

fometimes

sometimes in such numbers, that *Aurenge Zebe* used to send quarters as presents to all his great people *.

They are usually very gentle and tame, will feed readily, and lick the hands which give them food. In confinement they will eat oats, but prefer grafs and hay; and are very fond of wheaten bread. When thirsty, will drink two gallons at a time.

MANNERS.

They are said to be at times very vicious and fierce. When the males fight, they drop on their knees at a distance from one another, make their approaches in that attitude, and when they come near, spring and dart at each other. They will often, in a state of confinement, fall into that posture without doing any harm. They will, notwithstanding, attack mankind unprovoked. A laborer, who was looking over some pales which inclosed a few of them, was alarmed by one of the males flying at him like lightning; but he was saved by the intervention of the wood-work, which it broke to pieces, and at the same time one of its horns.

They have bred in *England*. They are supposed to go nine months with young, and have sometimes two at a birth. The young is of the color of a fawn. The dung is round and small, and comes away in quantities at a time, like that of deer.

Dama. *Plinii lib.* xi. c. 37.　　　　xxxiv.
Cemas. *Ælian. An. lib.* xiv. *c.* 14.　Antilope dama. *Pallas Miscel.* 5. *Spicil.*
Le Nanguer. *De Buffon*, xii. 213. *tab.*　8.

33. SWIFT.

A. with round horns, eight inches long, reverting at their ends: length of the animal three feet ten inches; height two feet eight inches: general color tawny: belly, lower part of

* *Bernier voy. Cachemire*, 47.

the

the fides, rump, and thighs, white : on the fore-part of the neck a white fpot : but this fpecies varies in color.

Inhabits *Senegal*; is eafily tamed; very fwift. *Ælian* compares its flight to the rapidity of a whirlwind.

34. RED.   Le Nagor. *De Buffon,* xii. 326. *tab.* xlvi. Antilope redunca. *Pallas Spicil.* 8.

A. with horns five inches and a half long; one or two flight rings at the bafe : ears much longer than the horns : length, four feet; height, two feet three inches : hair ftiff and bright : in all parts of a reddifh color; paleft on the cheft. Tail very fhort. Inhabits *Senegal*, and the *Cape*, where it is very frequent, and is a common food.

35. CINEREOUS.   Antilope Eleotragus. *Schreber.* cclxvi.

A. with horns, elegantly marked with fpiral wreaths. Head, hind part of, and fides of the neck, back, fides, fhoulders, and thighs, of a moft elegant greyifh afh color. Tail fhort, covered with longifh hair of the fame color. Front of the neck, breaft, belly, and legs, of a pure white.

An elegant fpecies, defcribed from Mr. *Schreber's* print; probably a native of *Africa*.

36. FOREST.   Le Bofbok. *Alamand Supplem.* V. 37. tab. xv. A. Sylvatica. *Gmelin. Lin.* 192.

A. with the head and upper part of the body dark brown, approaching about the head and under the neck to red. Belly and infide of the thighs and legs white. Rump marked

with

with fmall round fpots of pure white. Horns ten inches long, almoft ftrait, bending very flightly forward, and twifted fpirally for more than the lower half. Ears long and pointed. Tail fix inches long, and covered with long white hairs. Female hornlefs.

Length to the tail three feet fix.

Inhabits the forefts a hundred and fixty leagues beyond the *Cape*; are often difcovered by their voice, which refembles the barking of a dog.

This fhould be placed as the link between this clafs and the preceding.

SIZE.
PLACE.

Allamand Supplem. V. 34. tab. xiii.

37. RITBOK.

A. with horns one foot three inches long, bending forward, annulated half way up, very fharp pointed; their length in a ftrait line from bafe to point only ten. The whole upper part of the animal of an afh colored grey. Throat, belly, buttocks, and infide of the legs, white. Ears very long, white within, and near each is a bald fpot. Tail eleven inches long, flat and covered with long white hairs.

The length of this fpecies from nofe to tail is four feet five.

Inhabits the country a hundred leagues to the north of the *Cape* of *Good Hope*. Are numerous, but go in fmall herds, and fometimes only the male and female confort together. They frequent

both

SIZE.
PLACE.

both woods and fountains overgrown with reeds; from which the Dutch call them *Rietrheebok*, or *Roebuck* of the reeds. The females are hornlefs.

**** With twifted horns.

38. STRIPED.	Strepficeros. *Caii opufc.* 56. *Gefner quad.* 309. *Icon.* 31. Le Condcma. *De Buffon*, xii. 301. *tab.* xxxix. *vol.* xv. 142. Antilope Strepficeros. *Pallas Mifcel.* 9.	*Spicil.* 17. *Schreber* cclxvii. Cerf du Cap de Bonne efperance. *Hift. et Com. Acad. Palatin. tom.* i. 4°7. *Br. Muf. Afhm. Muf.* LEV. MUS.

HORNS.

A. with fmooth horns, twifted fpirally, compreffed fideways, with a ridge on one fide following the wreaths: confift of three bends: are fometimes four feet and a half long, meafured in a ftrait line *. Thofe which I examined, were three feet nine inches long; very clofe at their bafe, and two feet feven inches and a half diftant at their points, which are round and fharp. The horns are naturally of a dufky-color, and wrinkled; but are generally brought over highly polifhed. The FEMALES are deftitute of horns.

In the upper jaw a hard horny fubftance, difpofed in ridges.

SIZE.

Length of the animal nine feet; height, four: body long and flender: legs flender: face brown, marked with two white lines proceeding from the corner of each eye, and uniting above the nofe: the color in general of a reddifh caft, mixed with grey: from the tail, along the top of the back, to the fhoulders, is a

---

* *Journal hiftorique*, &c. p. 42. where there is a good figure of this animal.

white

Striped Antelope.—— N.°38

N.º 39 Common Antelope, & the Lyre Chelys

white ftripe: from this are feven others, four pointing towards the thighs, and three towards the belly: but I have obferved them to vary in number of ftripes. On the upper part of the neck is a fhort mane: beneath the neck, from the throat to the breaft, are fome long hairs hanging down: the breaft and belly are grey. Tail two feet long, brown above, white beneath, black at the end. STRIPES.

Inhabits the *Cape* of *Good Hope*, where it is called *Coedoes*. This name (perverted to that of *Coudous)* M. *de Buffon* has applied to the *Indian Antelope*, N°. 21. I believe *Kolben* means this, by his wild goat, ii. 115. tab. vi. It is faid to leap to a moft aftonifhing height *.

<table>
<tr><td>Strepficeros et Addax? *Plinii lib.* viii. c. 53.
Gazella Africana, the Antilope. *Raii fyn. quad.* 79.
Tragus Strepficeros. *Klein quad.* 18.
Capra Cervicapra. C. cornibus teretibus, dimidiato-annulatis, flexuofis contortis. *Lin. fyft.* 96.</td><td>L'Antelope. *De Buffon*, xii. 215. *tab.* xxxv. xxxvi.
Allamands De Buffon, v. 58. *tab.* v.
La Gazelle. *Briffon quad.* 44.
Antilope cervicapra. *Pallas Mifcel.* 9. *Spicil.* 18. *tab.* i. ii. *Br. Muf. Afhm. Muf.* LEV. MUS.</td><td>39. COMMON.</td></tr>
</table>

A• with upright horns, twifted fpirally, furrounded almoft to the top with prominent rings; about fixteen inches long, twelve inches diftance between point and point: in fize, rather lefs than the fallow-deer or buck: orbits white: white fpot on each fide of the forehead: color, brown mixed with red, and

* *Forfter's Voy.* i. 84.

dusky : the belly and inside of the thighs white : tail short, black above, white beneath. The females want horns.

Inhabits *Barbary*. The form of these horns, when on the scull, is not unlike that of the antient Lyre, to which *Pliny* compares those of his *Strepsiceros* *. The *Brachia*, or sides of that LYRES. instrument, were frequently made of the horns of animals, as appears from antient gems. *Montfaucon* has engraved several.

To convey the idea of their structure, I caused the figure of one to be engraved, taken from the fifth volume of the Philosophical Transactions abridged, tab. xiv. p. 474. I prefer this to many other figures, as the shell of a tortoise forms the base; which gave rise to the beautiful comment on this passage in *Horace*, by Doctor *Molyneux*.

> O Testudinis aureæ
> Dulcem quæ strepitum, PIERI temporas!
> O *mutis* quoque *piscibus*
> Donatura Cygni, si libeat, sonum.

The art of giving to dumb fishes the voice of a Swan, was thought a strange idea, till that gentleman pointed out that a Tortoise made part of the Lyre; which animal was by the antients ranked in the class of fish †: and even gave the name of χελυς to that species of musical instrument. *Horace* again invokes his lyre by an address to the Tortoise; which flings light on a seven-stringed one preserved in the supplement to *Montfaucon* ‡.

> Tuque Testudo resonare septem
> Callida nervis,
> Nec loquax olim neque grata.

---

* *Plinii hist. nat.* lib. xi. c. 37.   † *Plinii nat. hist.* lib. ix. c. x.
‡ iii. tab. 75. fig. 6.

α BROWN

α Brown. Lidmeé ? *Shaw's travels.*

Less than a Roebuck, horns like those of the last: face, back, and sides of a very deep brown, the last bordered with tawny: belly and inside of the legs white: above each hoof a black spot: tail black above, white beneath. Inhabits *Bengal:* possibly also *Barbary*, being nearer the size of the *Lidmeé* than any other.

β Smooth horned. *De Buffon*, xii. 217. *tab.* xxxvi. *fig.* 3.

In my cabinet is a pair of horns twisted like those of the preceding, but quite smooth and black: they are joined together in a parallel direction, the points turned different ways: when thus mounted, they are carried by the *Faquirs* in *India*, by way of weapon. See Mus. Lev. where weapons formed of the horns of the species N° 30 are preserved.

N 2 ***** With

⁎⁎⁎⁎⁎ with horns bending in the middle, and reverting forwards towards their end.

40. BARBARY.

Gazella Africana cornibus brevioribus, ab imo ad fummum fere annulatis, et circum medium inflexis. *Raii fyn. quad.* 80.

La Gazelle. *De Buffon*, xii. 201. *tab*. xxiii. La Gazelle d'Afrique. *Briffon quad.* 45. Capra Dorcas. *Lin. fyft.* 96. Antilope Dorcas. *Pallas Spicil.* xii. 11.

A with horns twelve inches long, round, inclining firſt back-
. wards, bending in the middle, and then reverting forwards at their ends, and annulated with about thirteen rings on
their lower part: upper ſide of the body reddiſh brown; lower
part and buttocks white : along the ſides the two colors are ſeparated from each other by a ſtrong duſky line : on each knee a
tuft of hair : the *Dorcas* of *Ælian*, *lib.* xiv. *c.* 14.

Inhabits *Barbary*, *Ægypt*, and the *Levant* : goes in large
flocks.

41. FLATHORNED.

Le Kevel. *De Buffon*, xii. 204. *tab*. xxiv.

Antilope Kevella. *Pallas Mifcel.* 7. Spi- cil. xi. 6. 8. 15.

A with horns ſhaped like thoſe of the laſt, but flatted on
. their ſides ; the rings more numerous, from fourteen to
eighteen : the ſize equal to a ſmall roebuck : in colors and marks
reſembles the preceding.

3

Inhabits

Inhabits *Senegal.* This, the *Barbary,* and *Harneffed,* have the fame manners and food; live in great flocks, are eafily tamed, and are excellent meat.

Either this animal, or one of thofe nearly allied to it, is found in abundance in the country on the eaft fide of the *Cafpian* fea: the *Perfian* name of it is *Dfhairan,* not *Ahu,* which *Kæmpfer,* by fome miftake, applies to it.

Antilope pygargus. *Pallas Spicil.* i. 10. & xii. 15. Lev. Mus.   42. WHITE-FACED.

A. with horns like thofe of the *Kevel,* fixteen inches long; five between tip and tip; annulated in the male, fmooth in the female : ears feven inches long : face, and fpace between the horns, of a pure white: cheeks and neck of a fine bright bay: back, of a cinereous brown, dafhed with red: along the middle, a dark lift: fides, flanks, and fhoulders, a deep brown; feparated from the belly by a broad band of darker color.

Belly and rump, and a fmall fpace above the tail, white.

Trunk of the tail feven inches long, covered with black coarfe hairs, which extend four inches beyond the end of the trunk: hoofs fhort.

In fize fuperior to the buck, or fallow deer. The length of the fpecimen in the LEVERIAN MUSEUM is five feet four inches : height three feet to the top of the fhoulders.   SIZE.

Inhabits the countries north of the Cape of *Good Hope.*   PLACE.

La

43. SPRINGER.   La Gazelle a bourſe ſur le dos. *Allemande*. Antilope Euchore *Forſter*, *Schreber*, cclxxii.

A with the face, cheeks, noſe, chin, throat, and part of the
under ſide of the neck, white: a duſky line paſſes from the
baſe of each horn, and beyond the eyes, to the corner of the
mouth.

Horns ſlender: annulated half way: twice contorted. Ears
very long, duſky.

Whole upper ſide of the neck, part of the lower, the back,
ſides, and outſide of the limbs, of a pale yellowiſh brown.
Darkeſt on the hind of the neck. Cheſt, belly, and inſide of the
limbs, white: the ſides and belly divided by a broad band of
cheſnut, which runs down part of the ſhoulders.

Tail reaches to the firſt joint of the leg. The upper part is
white: the lower black, and furniſhed with long hair. The
under ſide appears nearly naked. Buttocks are white; and from
the tail, half way up the back, is a ſtripe of white, expanſible at
pleaſure.

SIZE.   This elegant ſpecies weighs about fifty pounds, and is rather
leſſer than a roebuck.

PLACE.   Inhabits the Cape of *Good Hope*: called there the *Spring-bock*,
from the prodigious leaps it takes on the ſight of any body.
When alarmed, it has the power of expanding the white ſpace
about the tail into the form of a circle, which returns to its
linear form when the animal is tranquil.

They migrate annually from the interior parts in ſmall herds,
and continue in the neighborhood of the *Cape* for two or three
months:

months: then join companies, and go off in troops confifting of many thoufands, covering the great plains for feveral hours in their paffage. Are attended in their migrations by numbers of lions, hyænas, and other wild beafts, which make great deftruction among them. Are excellent eating, and, with other Antelopes, are the venifon of the *Cape*.

Mr. *Maffon* * informs us, that they alfo make periodical migrations, in feven or eight years, in herds of many hundred thoufands, from the north, as he fuppofes from the interior parts of *Terra de Natal*. They are compelled to it by the exceffive drought which happens in that region, when fometimes there does not fall a drop of rain for two or three years. Thefe animals in their courfe defolate *Caffraria*, fpreading over the whole country, and not leaving a blade of grafs. Lions attend them; where one of thofe beafts of prey are, his place is known by the vaft void vifible in the middle of the timorous herd. On its approach to the Cape, it is obferved that the avant guard is very fat, the centre lefs fo, and the rear guard almoft ftarved, being reduced to live on the roots of the plants devoured by thofe which went before; but on their return, *they* become the avant guard, and thrive in their turn on the renewed vegetation: while the former, now changed into the rear guard, are famifhed by being compelled to take up with the leavings of the others. Thefe animals are quite fearlefs, when affembled in fuch mighty armies, nor can a man pafs through unlefs he compels them to give way with a whip or ftick. When taken young they are eafily domefticated: the males are very wanton, and apt to butt at ftrangers with their horns.

* *Phil. Tranf* lxvi. 310.

Caprea

44. CHINESE. Caprea campeſtris gutturoſa. *Nov. Com.* 278, 290. *Le Brun,* i. 115.
Petrop. v. 347. *tab.* ix. Le Tzeiran *de* Antilope. *Bell's travels,* i. 311. 319.
Buffon, xii. 207. A. gutturoſa. *Pallas Spicil.* xii. 14. 46. *tab.*
Yellow Goat. *Du Halde China,* ii. 253, ii.

A. with horns about nine inches long, of a yellow color, opake, annulated almoſt to their ends, reclining backwards, diverging much at the upper part, with their points bending towards one another. Head rather thick. Noſe very blunt, and convex above. Ears ſmall, ſharp-pointed. On the middle of the neck is a great protuberance, occaſioned by the uncommon ſtructure of the windpipe. Tail not five inches long.

The hair on the approach of winter grows long, rough, and hoary; ſo that at a diſtance it appears almoſt white. In the beginning of *May,* the animal changes its coat for one very ſhort, cloſe, and tawny.

FEMALES. The females are hornleſs; but do not differ in color from the males.

Length of a male from noſe to tail about four feet and a half. Weight from eighty-one to ninety-eight pounds.

PLACE. Theſe animals abound in the country of the *Mongal Tartars,* and the deſerts between *Thibet* and *China,* and along the river *Amur* to the Eaſtern Sea. They are found alſo between the country of *Tangut* and the borders of *India.*

NAMES. The *Mongals* call them *Dſeren*; the *Chineſe, Hoang Yang,* and
MANNERS. *Whang Yang,* or Yellow Goats*. They are very ſwift, and take prodigious leaps, and when frightened will bound over three or four fathoms ſpace at one ſpring. Are very ſhy and timorous:

* *Du Halde,* ii. 253.

love

love dry and rocky plains: fhun water; nor will they go into it even to fave their lives, when driven by dogs or men to the brink of a river*. Are equally fearful of woods.

Go in fmall flocks in fpring and fummer: collect in great numbers in winter. They do not run confufedly, but in a file †, one after another; an old one leading the way. Seldom emit any voice. If taken young, are eafily tamed. Are objects of chace, being a great food among the *Tartars*. Their horns are an article of commerce, and in great requeft with the *Chinefe*. Thefe are the *Ablavos* ‡, which *Le Brun* met with by thoufands near lake *Baikal*, in the land of the *Burattes*.

A. Sub-gutturofa. *Act. Petr.* 1778. i. 251. tab. 9. 12. *Gmelin, Lin.* 186.

45. GUILDEN-STEDT'S.

A. with horns fhaped like the former, but of the length of thirteen inches: color of the body and outfides of the legs and thighs cinereous brown: tail fhort and full of hair: form of neck, breaft, and belly, white: fpace round the vent of the fame color. On the fore part of the neck is a protuberance, but leffer than that of the former. Knees tufted: fize of a roebuck: inhabits *Perfia*, between the *Cafpian* and *Euxine* feas: is gregarious: feeds chiefly on the *artemifia pontica*. The flefh delicious: the female brings forth in *May*, difcovered by that able traveller the late Mr. *Guildenftedt*.

PLACE.

* In my former edition I was mifled by *Gmelin* into a very different opinion.
† *Du Halde*, ii. 290.
‡ Doctor *Pallas*.

46. SCYTHIAN.

Colus. *Gesner quad.* 361.
Suhak. *Rzaczinski hist. Polon.* 224.
Ibex imberbis. *Nov. com. Petrop.* v.
tab. xix. vii. 39. xiv. 512.
Sayga. *Phil. Tr.* 1767. *p.* 344. *Bell's travels,* i. 43.
Capra Tatarica. C. cornibus teretibus

rectiusculis perfecté annulatis apice diaphanis gula imberbi. *Lin. syst.* 97.
Le Saiga. *de Buffon,* xii. 198. *tab.* xxii. *fig.* 2. Suppl. vi. 149.
Antilope Scythica. *Pallas spicil.* xii. 21. *tab.* i. *Faunul. sinens.* LEV. MUS.

HORNS.

A. with horns distant at the base, and with three curvatures; the last pointing inward. Stand a little reclining: the greatest part annulated: ends smooth. Color a pale yellow. Are semi-pellucid: length about eleven inches.

Head rather large. Nose in the live animal much arched and thick: very cartilaginous: divided lengthways by a small furrow: end as if truncated.

Ears small: irides of a yellowish brown. Neck slender: prominent about the throat. Knees guarded by tufts of hair.

The hair, during summer, is very short: grey mixed with yellow: below the knees darker. Space about the cheeks whitish: forehead and crown hoary, and covered with longer hairs. Under side of the neck and body white.

Winter coat long, rough, and hoary.

TAIL.

Tail four inches long: naked below; above cloathed with upright hairs, ending with a tuft.

SIZE.

Size of a fallow deer.

Females destitute of horns.

PLACE.

These animals inhabit all the deserts from the *Danube* and *Dnieper* to the river *Irtish,* but not beyond. Nor are they ever

seen

feen to the north of 54 or 55 degrees of latitude. They are found therefore in *Poland*, *Moldavia*, about Mount *Caucafus*, and the *Cafpian* Sea, ànd *Siberia*, in the dreary open deferts, where falt-fprings abound, feeding on the falt, the acrid and aromatic plants of thofe countries, and grow in the fummer-time very fat: but their flefh acquires a tafte difagreeable to many people, and is fcarcely eatable, until it is fuffered to grow cold after dreffing.

FOOD.

The females go with young the whole winter; and bring forth in the northern deferts in *May*. They have but one at a time: which is fingular, as the numbers of thefe animals are prodigious. The young are covered with a foft fleece, like new-dropt lambs, curled and waved.

They are regularly migratory. In the rutting-feafon, late in autumn, they collect in flocks of thoufands, and retire into the fouthern deferts. In the fpring they divide into little flocks, and return northward at the fame time as the wandering *Tartars* change their quarters.

MIGRATORY.

They very feldom feed alone; the males feeding promifcuoufly with the females and their young. They rarely lie down all at the fame time: but by a providential inftinct fome are always keeping watch: and when they are tired, they feemingly give notice to fuch which have taken their reft, who arife inftantly, and as it were relieve the centinels of the preceding hours. They thus often preferve themfelves from the attack of wolves, and from the furprize of the huntfmen*.

They are exceffively fwift, and will outrun the fwifteft horfe or gre-hound: yet partly through fear, for they are the moft timid

SWIFT.

* Doctor *Pallas.*

O 2

of

of animals, and partly by the ſhortneſs of their breath, they are **VERY TIMID.** very ſoon taken. If they are but bit by a dog, they inſtantly fall down, nor will they even offer to riſe. In running they ſeem to incline on one ſide, and their courſe is ſo rapid that their feet ſeem ſcarcely to touch the ground *.

SHORT-SIGHTED. They are during ſummer almoſt purblind; which is another cauſe of their deſtruction. This is cauſed by the heat of the ſun, and the ſplendor of the yellow deſerts they are ſo converſant in.

In a wild ſtate they ſeem to have no voice. When brought up tame, the young emit a ſhort ſort of bleating, like ſheep.

LIBIDINOUS. The males are moſt libidinous animals: the *Tartars*, who have ſufficient time to obſerve them, report that they will copulate twenty times together; and that this turn ariſes from their feeding on a certain herb, which has moſt invigorating powers.

When taken young, they may eaſily be made tame: but if caught when at full age, are ſo wild and ſo obſtinate as to refuſe all food. When they die, their noſes are quite flaccid.

CHACE. They are hunted for the ſake of their fleſh, horns, and ſkins, which are excellent for gloves, belts, &c. The huntſmen always approach them againſt the wind, leaſt they ſhould ſmell their enemy: they alſo avoid putting on red or white cloaths, or any colors which might attract their notice. They are either ſhot, or taken by dogs; or by the BLACK EAGLE †, which is trained to this ſpecies of falconry.

No animals are ſo ſubject to vary in their horns; but the color and clearneſs will always point out the animal to which they belong.

* Dr. *Cook's* travels, i. 317.   † *Br. Zool.* i. N° 2.

3

This

This probably was the animal called by *Strabo* Κολος *, found among the *Scythæ* and *Sarmatæ*, and an object of chace with the antient inhabitants. He fays it was of a fize between a ftag and a ram, and of a white color, and very fwift. He adds, that it drew up fo much water into its head, through its noftrils, as would ferve it for feveral days in the arid deferts: a fable naturally formed, in days of ignorance, from the inflated appearance of its nofe.

Le Corine. *de Buffon*, xii. 205. *tab.* xxvii. LEV. Mus †.

A. with very flender horns, fix inches long, furrounded with circular *rugæ:* ears large: lefs than a roebuck: on each fide of the face is a white line: beneath that is one of black: neck, body, and flanks, tawny: belly and infide of the thighs white: feparated from the fides by a dark line: on the knees is a tuft of hair.

Inhabits *Senegal.* Doctor *Pallas* doubts if this is not the female of the flat-horned, N° 32; but the form of the horns prevents my affent.

* *Lib.* vii. *p.* 480.   † A fine entire fpecimen.

Bubalus

48. CERVINE.   Bubalus. *Plinii lib.* viii. *c.* 15. βɤᵇαλⓈ-?   i. 205.
      Oppian *Cyneg.* ii. *Lin.* 300.   Le Bubale *de Buffon,* xii. 294. *tab.* xxxvii.
      Bufelaphus. *Gefner quad.* 121.   xxxviii.
      Capra Dorcas. *Lin. fyft.*   Antilope Bubalis. *Pallas fpicil.* xii. 16.
      Vache de Barbarie. *Memoire de L'acad.*   MUS. LEV.

A. with horns bending outward and backward, almoft clofe at their bafe, and diftant at their points; twifted and annulated; very ftrong and black; fome are above twenty inches long, and above eleven in girth at the bafe: head large, and like that of an ox: eyes placed very high, and near to the horns: the form of the body a mixture of the ftag and heifer: height to the top of the fhoulders four feet: the tail rather more than a foot long, afinine, and terminated with a tuft of hair: color, a reddifh brown: white about the rump, the inner fide of the thighs, and lower part of the belly: a dark fpace occupies the top of the back, the front of the upper part of the fore legs, and hinder part of the thighs.

Inhabits *Barbary*, and probably other parts of *Africa*, being alfo found towards the Cape of *Good Hope*. It is the *Bekker el wafh* of the *Arabs*, according to Dr. *Shaw*; who fays, that its young quickly grow tame, and herd with other cattle. Mr. *Forfkal* mentions it among the *Arabian* animals of an uncertain genus, by the name of *Bakar Uafch*. This is the *Bubalus* of the antients, not the *Buffalo*, as later writers have fuppofed. *Pliny* remarks an error of the fame kind in his days; fpeaking of the *Urus*, he fays, *Uros, quibus imperitum vulgus* bubalorum *nomen imponit, cum id gignat* Africa, *vituli potius cervive quadam fimilitudine.*

The

P. Mazell Sculp

Cervine Antelope.—N.°48.

The *Dutch* of the *Cape* call this species, *Hartebeeft*. They go in great herds; few only are folitary. Gallop feemingly with a heavy pace, yet go fwiftly. Drop on their knees to fight, like the *white-footed Antelope*, or *Nil-ghau*. The flefh is fine-grained, but dry *.

Le Koba. *De Buffon*, xii. 210. 267. *tab.*   cclxxvii.
xxxii. *fig.* 2.   Antelope Bubalis. *Pallas fpicil.* xii. 16.   49. SENEGAL.
Cerf qu'on nomment Temamaçama. *Seb.*   LEV. MUS.
Muf. i. 69. *tab.* xlii. *fig.* 4. *Schreber.*   Bucula cervina. *Caii* opufc. 63.

A. with horns almoft clofe at the bafe, a little above bending out greatly; then approach again towards the ends, and recede from each other towards the points, which bend backwards; the diftance in the middle fix inches and a half; above that four inches; at the points fix; length, feventeen inches; circumference at the bottom eight; furrounded with fifteen prominent rings; the ends fmooth and fharp: head large and clumfy, eighteen inches long: ears feven: head and body of a light reddifh brown: from the horns to the nofe along the face a ftripe of black: down the hind part of the neck a narrow black lift: rump, a dirty white: on each knee, and above the fetlock, a dufky mark: on the lower part of the ham and lower part of the fhoulders another: hoofs fmall: tail a foot long, covered with coarfe black hairs, which hang far beyond the end. Length of the whole fkin, which I bought at *Amfterdam*, feven feet.

Inhabits *Senegal*, where the *French* call it *La grande vache brune*. Certainly, neither the *Temamaçama* of *Hernandez*, nor even a native of *America*, as *Seba* afferts; nor yet to be made fynonymous with the former.

* *Sparman* in *Stockh.* Wettfk. Handl. 1779. p. 151.

Le

50. GAMBIAN. Le Kob, ou petite vache brune. *de Buffon,* xii. 210. 267. *tab.* xxxii. *fig.* 1.

A with horns thirteen inches long: five inches and a half round at the bottom: pretty close at the base and points; very distant in the middle. Surrounded with eight or nine rings: smooth at their upper part.

Inhabits *Senegal.*

N.º 50. . Nº 21..

Horns

XVII

Elk, or Moose Deer.____. N.º 51.

Horns upright, folid, branched, annually deciduous.   VII. DEER.
Eight cutting teeth in the lower jaw; none in the upper.

* With palmated horns.

Alce machlis, *Plinii lib.* viii. c. 15. *Gef-*   palmatis, caruncula gutturali, *Lin.*   51. ELK.
ner quad. i. 3. *Munfter Cofmog.* 883.   *fyft.* 92. Ælg. *Faun. Suec.* No. 39.
Cervus palmatus, Alce, Elant *Klein*   Los, *Rzaczinfki Polon.* 212.
quad. 24. *Ridinger wild Thiere.* 36.   C. cornibus ab imo ad fummum pal-
Allamand, xv. 50. *tab.* ii.   matis, *Briffon quad.* 6. *Faunul. Sinens.*
Elk, *Raii fyn. quad.* 86. *Scheffer Lapl.*   L'Elan, *de Buffon*, xii. 79. *tab.* vii. viii.
133. *Bell's trav.* i. 5. 215. 322.   *Br. Muf. Afh. Muf.* LEV. MUS.
Cervus Alces. C. cornibus acaulibus

D. with horns with fhort beams fpreading into large and   MALE.
broad palms, one fide of which is plain, the outmoft
furnifhed with feveral fharp fnags. No brow antlers*. The
largeft I have feen is in the houfe belonging to the *Hudfon Bay*

---

* In the *Britifh Mufeum* is a pair of Elk horns, which in all refpects refembles
the others, except that on the beam of each horn, about four inches from the
bafe, is a branch, round and trifurcated: very different from a brow-antler. It
is the only one of the kind I ever faw; fo, probably, is a mere accident; for
neither the many *European* Elks horns, or the feveral pair of *American* Elk or
Moofe, I have examined, are furnifhed with brow-antlers. Thofe in queftion
feem to be the very pair which Mr. *Dale* defcribes and figures, *Phil. Tranf.*
abridg. ix. 85. *tab.* 6. *fig.* 50.

VOL. I.   P   company;

company; weigh'd 56. lb; length 32 inches; between tip and tip, 34; breadth of the palm 13 ½. There is in the same place an excellent picture of an Elk, which was killed in the presence of *Charles* XI. of *Sweden*, and which weighed 1229 lb. The length of one killed on the *Altaic* mountains in *Sibiria*, from nose to tail, was eight feet ten inches, *Paris* measure. The height before, five feet six; behind, about two inches more. The full length of the head two feet five; yet this was not one of the largest. The tail was only two inches and one-third. It is a very deformed and seemingly disproportioned beast.

FEMALE.　　A young female of about a year old, was to the top of the withers five feet high, or fifteen hands; the head alone two feet long; length of the whole animal, from nose to tail, about seven feet: the neck much shorter than the head, with a short thick upright mane, of a light brown-color. The eyes small: the ears one foot long, very broad and slouching: nostrils very large: the upper lip square, hangs greatly over the lower, and has a deep sulcus in the middle, so as to appear almost bifid: nose very broad: under the throat a small excrescence, from whence hung a long tuft of coarse black hair: the withers very high: fore legs three feet three inches long; from the bottom of the hoof to the end of the *tibia* two feet four inches: the hind legs much shorter than the fore legs: hoofs very much cloven: tail very short; dusky above, white beneath: color of the body in general a hoary black; but more grey about the face than any where else. This was living at the Marquis of *Rockingham*'s house, at *Parson's-green*. It seemed a mild animal; was uneasy and restless at our presence, and made a plaintive noise. This

was

was brought from *North America*, and was called the *Moose Deer* *.

A male of this species, and the horns of others, having been brought over of late years, prove this, on comparison with the horns of the *European* Elk, to be the same animal. But the account that *Joſſelyn* † gives of the size of the *American Moose* has all the appearance of being greatly exaggerated; asserting, that some are found twelve feet or thirty-three hands high. But *Charlevoix, Dierville,* and *Leſcarbot* ‡, with greater appearance of probability, make it the size of a horse, or an *Auvergne* mule, which is a very large species; and the informations also that I have received from eye-witnesses, make its height from fifteen to seventeen hands. The writers who speak of the *European* kind, confine its bulk to that of a horse. Those who speak of the gigantic Moose, say, their horns are six feet high; *Joſſelyn* makes the extent from tip to tip to be two fathom; and *La Hontan* ‖, from hearsay, pretends, that they weigh from 300 to 400 lb.; notwithstanding he says, that the animal which is to carry them is no larger than a horse. Thus these writers vary from each other, and often are not consistent with themselves. It seems then that *Joſſelyn* has been too credulous, and takes his evidence from huntsmen or *Indians*, who were fond of the marvellous; for it does not appear that he had seen it. The only

MOOSE AND ELK THE SAME ANIMAL.

* From *Muſu,* which in the *Algonkin* language signifies an Elk. *Vide Kalm iter.* vol. iii. 510. *Germ. ed. De Laet.* 73. *Purchas Pilgr.* iv. 1831.

† *Joſſelyn's voy. New Engl.* 88. *New Engl. rarities,* 19.

‡ *Charlevoix hiſt. nouvelle France,* v. 185. *Dierville voy. de L'Acadie,* 122. *Leſcarbot hiſt. nouv. France,* 810. The *French* call this animal, *Original.*

‖ *Voy. N. America,* i. 57.

P 2

thing

thing certain is, that the Elk is common to both continents; and that the *American*, having larger forefts to range in, and more luxuriant food, grows to a larger fize than the *European*.

In *America* they are found, tho' rarely, in the back parts of *New England*; in the peninfula of *Nova Scotia*, and in *Canada*; and in the country round the great lakes, almoft as low fouth as the *Ohio*. In *Europe* they inhabit *Lapland*, *Norway*, *Sweden*, and *Ruffia*; in *Afia*, the N. E. parts of *Tartary* and *Siberia*; but in each of thofe continents inhabit only parts, where cold reigns with the utmoft rigour during part of the year.

They live amidft the forefts, for the conveniency of browzing the boughs of trees: by reafon of the great length of their legs, and the fhortnefs of their neck, which prevent them from grazing with any fort of eafe, they often feed on water-plants, which they can readily get at by wading; and *M. Sarrafin** fays, they are fo fond of the *Anagyris fœtida*, or ftinking bean trefoil, as to dig for it with their feet, when covered with fnow.

They have a fingular gait; their pace is a high fhambling trot, but they go with vaft fwiftnefs; in old times thefe animals were made ufe of in *Sweden* to draw fledges; but as they were frequently acceffary to the efcape of murderers and other criminals, the ufe was prohibited under great penalties. In paffing thro' thick woods, they carry their heads horizontally, to prevent their horns being entangled in the branches. In their common walk they raife their fore-feet very high; that which I faw ftepped over a rail near a yard high with great eafe.

They are very inoffenfive animals, except when wounded, or in

* *Martyn's abridg. mem. and hift. Acad.* iv. 253.

the

the rutting-feafon, when they become very furious, and at that time fwim from ifle to ifle, in purfuit of the females. They ftrike with both horns and hoofs. Are hunted in *Canada* during winter, when they fink fo deep in the fnow as to become an eafy prey: when firft unharbored, fquat with their hind parts, make water, and then go off in a moft rapid trot: during their former attitude, the hunter ufually directs his fhot.

The flefh is much commended for being light and nourifhing, but the nofe is reckoned the greateft delicacy in all *Canada*: the tongues are excellent, and are frequently brought here from *Ruffia*: the fkin makes excellent buff leather*; *Linnæus* fays, it will turn a mufket-ball: the hair which is on the neck, withers, and hams, of the full-grown Elk, is of great length, and very elaftic; is ufed to make matreffes. The hoofs were fuppofed to have great virtues, in curing epilepfies. It was pretended that the Elk, being fubject to that difeafe, cured itfelf by fcratching its ears with its hoof.

The Elk was known to the *Romans* by the name of *Alce* and *Machlis*: they believed that it had no joints in its legs; and, from the great fize of the upper lip, imagined it could not graze without going backward.

Before I quit this fubject, it will be proper to take fome notice of the enormous horns that are fo often found foffil in *Ireland*, and which have always been attributed to the Moofe Deer: I mean the Moofe Deer of *Joffelyn*; for no other animal could poffibly be fuppofed to carry fo gigantic a head. Thefe horns differ

* Numbers of the *American* Elk-fkins are fent from hence to *Bayonne*, where they are dreffed, and fold to the *Gallegos*, who make buff waiftcoats of them.

very

very much from thofe of the *European* or *American* Elk; the beam, or part between the bafe and the palm, is vaftly longer: each is furnifhed with a large and palmated brow antler, and the fnags on the upper palms are longer. The meafurements of a pair of thefe horns are as follow: from the infertion to the tips, five feet five inches; the brow antlers eleven inches; the broadeft part of the palm, eighteen; diftance between tip and tip, feven feet nine: but thefe are fmall in comparifon of others that have been found in the fame kingdom. Mr. *Wright*, in his *Louthiana, tab.* xxii. book III. gives the figure of one that was eight feet long, and fourteen between point and point. Thefe horns are frequent in our *Mufeums*, and at gentlemen's houfes in *Ireland:* but the Zoologift is ftill at a lofs for the recent animal. I was once informed by a gentleman long refident in *Hudfon's Bay*, that the *Indians* fpeak of a beaft of the Moofe kind (which they call *Wafkeffer*) but far fuperior in fize to the common one, which they fay is found 7 or 800 miles S. W. of *York Fort*. If fuch an animal exifted, with horns of the dimenfions juft mentioned, and of proportionable dimenfions in other parts, there was a chance of feeing *Joffelyn's* account verified: for if our largeft elks of feventeen hands high carry horns of fcarcely three feet in length, we may very well allow the animal to be thirty-three hands high which is to fupport horns of 3 or 400 lb. weight. But from later enquiries, I find that the *Wafkeffer* of the *Indians* is no other than the animal we have been defcribing.

Tarandus?

Rain Deer. ____ 52.

Tarandus? *Plinii lib.* viii. c. 34.
Le Rangier ou Ranglier. *Gaſton de Foix chez du Fouilloux,* 98.
Tarandus, Rangifer. *Gefner quad.* 839, 840. *Icon. quad.* 57, 58.
Cervus mirabilis. *Jonſton quad. Munſter Coſmog.* 1054.
Macarib, Caribo, Pohano. *Joſſelyn's New England rarities,* 20.
Cervus rangifer. *Raii ſyn. quad.* 88.
Rennthier. *Klein quad.* 23. *Ridinger wild Thiere.* 35.
C. Tarandus. C. cornibus ramoſis re curvatis teretibus, ſummitatibus palmatis. *Lin. ſyſt.* 93. *Schreber,* tab. ccxlviii. A. B. C.

Rhen. *Faun. Suec. No.* 41. *Amæn. Acad.* iv. 144.
Le Renne. *de Buffon,* xii. 79. *tab.* x. xi. xii. *Allamand,* xv. 50. *tab.* iii. *Briſſon quad.* 63.
Reindeer. *Scheffer Suppl.* 82. 129. *Le Brun's travels,* i. 10, 11. *Œuvres de Maupertuis,* iii. 198. *Voyage d'Outhier,* 141. *Hiſt. Kamtſchatka,* 228. *Bell's travels,* i. 213. *Martin's Spitzberg,* 99. *Crantz Greenl.* i. 70. *Egede Greenl.* 60. *Dobbs's Hudſon's bay,* 20. 22. *voy. Hudſ. bay.* ii. 17. 18.
Le Caribou. *Charlevoix hiſt. nouv. France,* v. 190. *Br. Muſ. Aſhm. Muſ.* LEV. MUS.

D. with large but ſlender horns, bending forwards; the top palmated, with brow antlers broad and palmated: horns on both ſexes; thoſe of the female leſs, and with fewer branches. A pair from *Greenland* was three feet nine inches long; two feet ſix from tip to tip; weighed 9 lb. 12 oz. Height of a full-grown Rein, four feet ſix. Space round the eyes always black. When it firſt ſheds its coat, the hairs are of a browniſh aſh-color; after that, changes to white; the hairs are very cloſely ſet together; along the fore-part of the neck are very long and pendent: hoofs large and concave; tail ſhort.

Inhabits farther north than any other hoofed quadruped. In *America,* it is found in *Spitzbergen,* and *Greenland,* but not further ſouth than *Canada*; in *Europe,* abounds in *Samoidea, Lapland, Norway*; in *Aſia,* the north coaſt, as far as *Kamtſchatka,* and the inland parts as low as *Siberia.* Found in all theſe places in a ſtate of nature; is domeſticated only by the *Laplanders, Samoides* and *Kamtſchatkans*; is to the firſt the ſubſtitute

PLACE.

of

of the horfe, the cow, the goat, and the fheep; and is their only wealth. The milk of the Rein affords them cheefe; the flefh, food; the fkin, cloathing; the tendons, bowftrings; and when fplit, thread; the horns, glue; the bones, fpoons. During the winter it fupplies the want of a horfe, and draws their fledges with amazing fwiftnefs over the frozen lakes and rivers; or over the fnow, which at that feafon covers the whole country. In running makes a great clatter with the collifion of the fpurious hoofs, which are large and loofe. It does not gallop in the manner reprefented in the figure of it in my firft edition, or as reprefented by Mr. *Ridinger*, in the 35th plate of his *Wilden Thiere*; but has a rapid running pace. A rich *Laplander* is poffeffed of a herd of a thoufand Reins. In autumn they feek the higheft hills, to avoid the *Lapland Gadfly**, which at that time depofits its eggs in their fkin; and is the peft of thefe animals, for numbers die that are thus vifited. The moment a fingle fly appears, the whole herd inftantly perceives it: they fling up their heads, tofs about their horns, and at once attempt to fly for fhelter amidft the fnows on the loftieft *Alps*. In fummer they feed on feveral plants; but during winter, on the rein-liverwort †, which lies far beneath the fnow; which they remove with their feet and palmated brow antlers, in order to get at their beloved food.

My very worthy friend, the late Doctor *Ramfay*, profeffor of Natural Hiftory in *Edinburgh*, affured me, that the horns of this fpecies were found foffil, in 1775, in a marle-pit, five feet below the furface, near *Craigton*, in the fhire of *Linlithgow*. They live only fixteen years.

* Œftrus Tarandi. *Faun. Suec. No.* 1731. *Flor. Lap.* 360,
† Lichen rangiferinus, *fp. pl.* ii. 1620. *Flor. Lap.* 331.

The

The horns vary in fize, and a little in form: one at Mr. *John Hunter*'s, has two broad four-furcated branches over the brow antlers, bending a little inwards: the whole was ftronger and broader, in proportion to the length, than common, and of a dull deep yellow color. Thefe are faid to be the horns of the female.

Προξ. *Arift. hift. An. lib.* ii. c. 14.
Platyceros. *Plinii lib.* xi. c. 38. *Oppian Cyneg. lib.* ii. lin. 293.
Platogna. *Belon obf.* 55.
Dama vulgaris five recentiorum. *Gefner quad.* 307.
Daniel. *Rzaczinfki Polon.* 217.
Cervus Platyceros, Fallow Deer. *Raii syn. quad.* 85.
Cervus palmatus. **Dam-tanhirfch.** *Klein quad.* 25.

Cervus dama. C. cornibus ramofis re-curvatis compreffis: fummitate pal-mata. *Lin. fyft.* 93. *Haffelquift, itin.* 290.
Dof, Dofhiort. *Faun. fuec. No.* 42.
Le Dain. *de Buffon,* vi. 161. *tab.* xxvii. *Briffon quad.* 62.
Buck. *Br. Zool.* i. 34. *Pontop. Norway,* ii. 9. *Du Halde China,* i. 315. *Faunul. finens.* LEV. MUS.

53. FALLOW.

D. with horns palmated at their ends and pointing a little for-ward, and branched on the hinder fide; two fharp and flender brow antlers, and above them two fmall flender branches. Color of this deer various, reddifh, deep brown, white, fpotted.

Not fo univerfal as the Stag; rare in *France* and *Germany.* Found wild in the woods of *Lithuania** and *Moldavia†*, in *Greece*, the *Holy Land*, and the north of *China*. In great abun-dance in *England*; but, except on a few chafes, at prefent confined in parks. M. *de Buffon* fays, that the fallow-deer of *Spain* are almoft as large as ftags. None originally in *America*. What are

PLACE.

* *Rzaczinfki.* † Doctor *Pallas.*

improperly

improperly called by that name will be deſcribed hereafter. . Are eaſily tamed : during rutting time, will conteſt with each other for their miſtreſs ; but are leſs fierce than the ſtag: during that ſeaſon, will form a hole in the ground, make the female lie down in it, and then often walk round and ſmell at her.

** With rounded horns.

54. STAG. | Cervus. *Plinii lib.* viii. *c.* 32. *Geſner quad.* 326.
Jelen. *Rzaczinſki Polon.* 216.
Red Deer, Stag, or Hart. *Raii ſyn. quad.* 84.
Cervus nobilis. Hirſch. *Klein quad.* 23.
C. Elaphus. C. cornibus ramoſis teretibus recurvatis. *Lin. ſyſt.* 93. Hiort.

Kron-hiort. *Faun. ſuec.* No. 4.
Le Cerf *De Buffon*, vi. 63. *tab.* ix. x. *Briſſon quad.* 58.
Stag, or Red Deer. *Br. Zool.* i. 34. *Shaw's travels*, 243.
Cateſby Carolin. Acc. xxviii. *Lawſon Carolin.* 123. *Faunul. ſinenſ.* LEV. MUS.

D. with long upright horns, much branched : ſlender and ſharp brow antlers. Color of the ſtag generally a reddiſh brown, with ſome black about the face, and a black liſt down the hind-part of the neck and between the ſhoulders. Grows to a large ſize; one killed in the county of *Aberdeen* weighed 18 ſtone *Scots*, or 314 lb. Horns of the *American* ſtags ſometimes weigh 30 lb. and are above four feet high.

PLACE. Common to *Europe*, *Barbary*, north of *Aſia*, and *North America*. Numerous in the ſouthern track of *Siberia*, where it grows to a monſtrous ſize. Extirpated in *Ruſſia*. Are ſtill found in a ſtate of nature in the highlands of *Scotland*. Lives in herds : one male generally ſupreme in each herd. Furious and dangerous in rutting-time. Seeks the female with a violent braying. Rutting-

ſeaſon

feafon in *Auguft*. Begins to fhed its horns the latter end of *February*, or beginning of *March:* recovers them entirely in *July*. Fond of the found of the pipe; will ftand and liften attentively. *Waller*, in his ode to Lady *Ifabella* on her playing on the lute, has this allufion to the fondnefs of the animal for mufic:

> Here Love takes ftand, and, while fhe charms the ear,
> Empties his quiver on the liftening deer.

PLAYFORD, in his introduction to mufic, has the following curious paffage to this purpofe: "Myfelf," fays he, "as I travelled "fome years fince near *Royfton*, met a herd of ftags, about "twenty, on the road, following a bag-pipe and violin; which, "while the mufic played, they went forward, when it ceafed, "they all ftood ftill; and in this manner they were brought out "of *Yorkfhire* to *Hampton Court* *."

The account of the *Cervina Senectus* †, or vaft longevity of the ftag, fabulous. Hinds go with young above eight months, bring one at a time, feldom two: fecure the young from the ftag, who would deftroy it. Flefh of thefe animals coarfe and rank: fkin ufeful for many purpofes: from the horns is extracted the celebrated fpirit of *hartfhorn*; but the horns of all other deer yield the fame falt. The *Hippelaphus* ‡ of the antients, only a large race of ftags, with longer hair on the neck, giving it the appearance of a mane. This is diftinguifhed by the *French* with

---

* STILLINGFLEET's Principles and Power of Harmony, 183.

† *Juvenal*, Sat. xiv. 251. *Plinii lib.* viii. *c.* 33, fpeaks of fome that were taken about 100 years before his time, with golden collars on their necks, which had been put on them by *Alexander the Great*.

‡ *Ariftot. Hift. An.* lib. ii. *c.* 1.

Q 2

the

the title of *Cerf d' Ardenne:* by the *Germans,* with that of *Brand‑hirtz.* Under the same variety may be also brought the *Trage‑laphus* of *Gesner,* so called from being more hairy than common *.

Le *Cerf de Corse* of M. *de Buffon,* vi. is the left species, of a deep brown color. *Vide* p. 95. *tab.* xi. This may be the same as the small kind of stag, rather larger than the fallow-deer, which Dr. *Shaw* says is found in *Barbary,* whose female the *Moors* call in derision *Fortass,* or Scald-head, from having no horns †.

Du Halde, i. 122. speaks of a small sort of stag, found in *Sun‑nan,* a province of *China,* not bigger than a common dog.

55. VIRGINIAN.	Fallow-deer. *Lawson Carol.* 123. *Catesby,* *Acc.* xxviii. *du Pratz,* ii. 50. Dama Virginiana. *Raii syn. quad.* 86. *Ph.*	*Tr. abridg.* ix. 86. *Br. Mus. Ashm.* *Mus.* LEV. MUS.

D. with slender horns, bending very much forward : nu‑merous branches on the interior sides; no brow ant‑lers: about the size of the *English* fallow-deer: of a light color, a cinereous brown : tail ten inches long. A quite distinct species, and peculiar to *America.*

MANNERS. Are found in vast herds. Those near the shores are lean and bad, and subject to worms in their heads and throats. Are very restless; always in motion; not fierce: their flesh dry; but of the utmost importance to the *Indians,* who dry it for their winter provision. The skins a great article of commerce, vast numbers annually imported from our colonies. Feed during hard winters

* *Gesner quad.* 296. Distinct from the Tragelaphus *Caii.*
† *Travels,* 243.

on the mofs which hangs in long ftrings from the *American* trees, in the northern parts. Are very eafily made tame, fo as to return to their mafter at night, after feeding all day in the woods. Thefe, not the Roe, as quoted by M. *de Buffon* *, are intended by *Kalm* †, and probably by M. *Fontannette.*

Axis. *Plinii lib.* viii. *c.* 21.   *Belon obf.* L'Axis. *de Buffon,* xi. 397. *tab.* xxxviii.   56. SPOTTED.
119. (fæm.) *Raii fyn. quad.* 89.   xxxix.   AXIS.
Speckled Deer. *Nieuhoff voy.* 262.

D. with flender trifurcated horns; the firft branch near the bafe; the fecond near the top; each pointing upwards: fize of the fallow-deer: of a light red color: the body beautifully marked with white fpots: along the lower part of the fides, next the belly, is a line of white: the tail long; as that of a fallow-deer; red above, white beneath.

Common on the banks of the *Ganges,* and in the ifle of *Ceylon. Pliny* defcribes them well among the animals of *India,* and adds that they were facred to *Bacchus.* They will bear our climate; and have bred in the Prince of *Orange's* menagery near the *Hague:* are very tame: have the fenfe of fmelling very exquifite: readily eat bread, but will refufe a piece that has been breathed on: many other animals of this, the antelope, and goat kind, will do the fame.

---

* *de Buffon,* Supplem. iii. 125.   † Travels, i. 209.

D. with

D. with rough and ftrong horns, trifurcated. The color of the hair is the fame with the former. Is of a middle fize between the fpotted and the great, or equal to that of our ftag; and is never fpotted; but fometimes varies to white, and is reckoned a great rarity.

Inhabits the dry hilly forefts of *Ceylon, Borneo, Celebes*, and *Java*, in herds of hundreds. In *Java* and *Celebes* they grow very fat: in thofe two iflands are great hunting-matches, and multitudes are killed at a time. The flefh is cut into fmall pieces, and dried in the fun, and falted for ufe.

In the *Britifh Mufeum* is a pair of large horns, of the fame fhape with the former, and, like them, trifurcated; are very thick, ftrong, and rugged; of a whitifh color; two feet nine inches long; two feet four inches between tip and tip.

Thefe probably came from *Borneo* or *Ceylon*. Mr. *Loten* having informed me of a fpecies of ftag in thofe iflands as tall as a horfe, and with horns three-forked. They are of a reddifh-brown color. The *Dutch* call them *Elanden*, or *Elks*. In *Borneo*, they are found in low marfhy places, for which reafon they are there called, in the *Javan* and *Malayan* language, *Mejangan Banjoe*, or water ftags.

The fpecies of Deer, probably one of the three laft, are found in *Mindanao, Gilolo, Mandioly, Batchian*, and all the *Papuas* iflands. Oxen, buffaloes, goats, hogs, dogs, cats, and rats are alfo found there, but no kind of beafts of prey. In *New Guinea*, all thofe kinds of quadrupeds ceafe, except the dog and hog.

D. with

P. Mazell Sculp.

Porcine Deer. Nᵒ 59.

D. with slender trifurcated horns, thirteen inches long; six inches distant at the base: head ten inches and a half long: body, from the tip of the nose to the tail, three feet six inches: height, from the shoulders to the hoof, two feet two inches; and about two inches higher behind: length of the tail eight inches: body thick and clumsy: legs fine and slender: color on the upper part of the neck, body, and sides, brown; belly and rump, of a lighter color.

    In possession of the late Lord *Clive*, brought from *Bengal*; called, from the thickness of their body, Hog Deer. The same species is also found in *Borneo*. They are taken in square pit-falls, about four feet deep, covered with some slight materials. Of their feet, as well as those of the lesser species of Musks and Antelopes, are made tobacco-stoppers.

D. with three longitudinal ribs extending from the horns to the eyes. Horns placed on a boney process, like a pedestal, elevated three inches above the scull, and covered with hair. The horns trifurcated; the upper fork hooked. From each of the upper jaws hangs a tusk.

    In size somewhat less than the *English* roe-buck, but of the shape of the Porcine deer. They live only in families. Inhabit *Java* and *Ceylon*; where they are called in the *Malaye* tongue *Kidang*, and by the *Javans*, *Munt-jak*: are common, and esteemed for the delicacy of their flesh.

The

59. PORCINE.

PLACE.

60. RIB-FACED.

SIZE.

The pedeſtals or pillars on which the horns ſtand, grow thicker as the deer advances in age; the margin alſo ſwells out around; ſo that if the horns are forced off the. pedeſtals, the ſurface of the laſt have the appearance of a roſe.

61 Roe.

Caprea. *Plinii lib.* xi. *c.* 37.
Caprea, capreolus, Dorcas. *Geſner quad.* 296.
Sarn. *Rzaczinſki Polon.* 27.
Cervus minimus. *Klein quad.* 24.
Cervus capreolus. C. cornibus ramoſis teretibus erectis, ſummitate bifida.

Lin. *ſyſt.* 94. Radjur. *Faun. ſuec,* No. 43.
Le Chevreuil. *de Buffon,* vi. 289. *tab.* xxxii. xxxiii. *Briſſon quad.* 61. *Charlevoix, N. Franc.* v. 195.
Roebuck. *Br. Zool.* i. 139, 200. *Br. Muſ.* *Aſh. Muſ.* Lev. Mus.

D. with ſtrong upright rugged trifurcated horns, from ſix to eight inches long: length,, from noſe to tail, three feet nine inches: height before, two feet three inches: behind, two feet ſeven inches: tail, one inch: weight of a full-grown buck near 60 lb. Hair in ſummer very ſhort and ſmooth; ends of the hairs deep red, bottoms dark grey: in winter very long, and hoary at the tips, except on the back, where it is often very dark: the legs ſlender; and below the firſt joint of the hind legs is a tuft of long hair: rump, and under ſide of the tail, white.

Place.

Inhabits moſt parts of *Europe,* as far north as *Norway:* is not found in *Africa.* Uncertain whether this kind is in *N. America,* notwithſtanding it is mentioned by *Charlevoix:* being unnoticed by *Lawſon, Cateſby, Kalm,* and *Du Pratz.* Frequent in the wooded parts of the highlands of *Scotland,* but, at preſent, in no other part of *Great Britain.*

Food.

Fond of mountanous wooded countries; brouzes very much, and

and during winter eats the young fhoots of fir and beech: is very active; lives in fmall families: brings two young at a time; conceals them from the buck. The flefh delicate, but never fat.

Cervus Pygargus. *Pallas Itin.* i. 453. C. Aha. S. Gmelin iter, iii. 496. *Gmelin, Lin.* **62. TAIL-LESS.**
175. *Schreber,* tab. ccliii.

D. with trifurcated horns like the former, very rugged at the bafe. The hairs of the eye-lids, and about the orbits, long and black. The infide of the ears covered with a very thick fur; nofe and fides of the under lip black: its tip white. No tail; only a broad cutaneous excrefcence above the anus.

Color of a roe-buck. About the buttocks is a great bed of a fnowy whitenefs, extending to the back.

Its whole coat excefsively thick; and in the fpring quite rough and erect.

Larger than the *European* kind*. Very common in all the   **SIZE.** temperate parts of *Ruffia* and *Siberia,* efpecially the fhrubby   **PLACE.** mountanous tracts beyond the *Volga,* and in the mountains of *Hyrcania.* But it does not extend to the N. E. of *Siberia.*

At approach of winter defcends into the open plains, and the hair in that feafon affumes a hoary appearance.

The *Perfians* call this animal, *Abu†.* The *Tartars* name it the   **NAMES.** *Saiga,* which properly fignifies the *roe-buck*; and is now adopted for the *Scythian* antelope by the inhabitants of the *Ruffian* empire ‡.

* Dr. *Pallas.* MS. The Roe-buck, *Bell's* Travels, i. 201, and Faunul. Sinens. of *Ofbeck,* may be of this kind.
† *Pallas,* Spicil. Zool. xii. 7.   ‡ The fame, 34.

63. MEXICAN.   Teutlalmaçame. *Hernandez An. Mexic.*   Biche des bois. *Barrere France Æquin.*
                    324.   151.
                    Caguaca-apara? *Marcgrave Brasil.* 235.   Chevreuil *d'Amerique, de Buffon,* vi. 210.
                    Piso Brasil. 97.   243. *tab.* xxxvii.
                    Baieu. *Bancroft Guiana,* 122.   Le Cariacou? *de Buffon,* xii. 324. 347.
                    Cervus major, corniculis brevissimis.   *tab.* xliv.

D. with strong thick rugged horns, bending forward; ten inches long; nine between point and point; trifurcated in the upper part: one erect snag about two inches above the base: by accident subject to vary in the number of branches: head large: neck thick: eyes large, and bright: about the size of the *European* Roe: color of the hair reddish: when young, spotted with white.

Inhabits *Mexico, Guiana,* and *Brasil;* not only the internal parts of the country, but even the borders of the plantations: the flesh inferior to that of *European* venison. A species very distinct from the Roe of the old continent. Perhaps this is the *wild goat* (as *Bossu* * calls it) which he says is plentiful in *Louisiana,* whose female has two *cornichons* or snags to its horns.

The *Squinaton,* or more properly the *Scenoontung,* an inhabitant of the countries west of *Hudson's Bay,* is another obscure animal, said to be less than a Buck and larger than a Roe, with finer legs and sharper head. An accurate account of the hoofed quadrupeds of the new continent, is among the *desiderata* of the Zoologist.

In the *Museum* of the Royal Society is a pair of horns of some animal of the Roebuck kind, styled by *Grew* †, horns of the *In-*

* *Travels.* i. 350.   † *Rarities,* 24.

dian

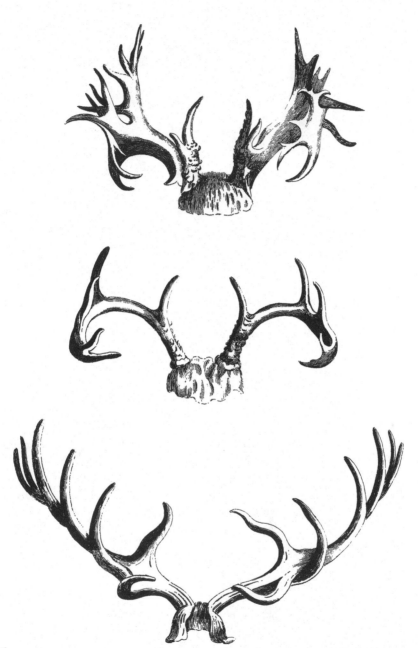

XX

Fossil Horns. p. 109. Horns of the Virginian Deer. p. 116.
Horns of the Mexican Deer. p. 122.

dian Roebuck: they are fixteen inches long, and the fame between tip and tip; are very thick, ftrong, and rugged; near the
bafe of each is an upright forked branch; the ends bend forward, divide into two branches, each furnifhed with numerous
fnags.

Cervus Guineenfis. C. grifeus fubtus nigricans. *Muf. Fr. Ad.* 12: *Lin.*
fyft. 94.

64. GREY.

AN obfcure fpecies, doubtful whether a Deer, a Mufk, or female Antelope; for the horns were wanting in the animal
defcribed by *Linnæus.*

Size of a cat; of a grey color: between the ears a line of
black: a large black fpot above the eyes: on each fide the
throat a line of the fame-color pointing downwards: the middle
of the breaft black: the fore legs and fides of the belly, as far
as the hams, marked with black: ears rather long: under fide
of the tail black.

R 2

** Without

** Without horns.

VIII. MUSK.   Two long tuſks in the upper jaw.
               Eight ſmall cutting teeth in the lower jaw; none in the upper.

65. TIBET.   Capreolus Moſchi. *Geſner quad.* 695.   *Du halde China,* i. 63. 324. *Grew's Mu-*
               Animal Moſchiferum. *Raii ſyn. quad.*   *ſeum,* 21.
               127. *Schrockius hiſt. Moſchi,* 1. *tab.* i.   Moſchus Moſchiferus. M. folliculo um-
               Animal Moſchiferum, Kabarga. *Nov.*   bilicali. *Lin. ſyſt.* 91.
               com. Petrop. iv. 393.   Tragulus, ſp. 5. Le Muſc. *Briſſon quad.*
               Muſk animal. *Tavernier's trav.* ii. 153.   67. *Klein quad.* 18.
               Le Brun's trav. i. 116. *Bell's trav.*   Le Muſc. *De Buffon,* xii. 361. *Faunul.*
               i. 249. ii. 88. *Strahlenberg,* 339.   *ſinenſ.* LEV. MUS.

M. of the form of a roebuck: length three feet three inches; from the top of the ſhoulders to the ſoles of the feet, two feet three inches. From the top of the haunches to thoſe of the hind feet, two feet nine inches.

Upper jaw much longer than the lower; on each ſide a ſlender tuſk, near two inches long, very ſhort on the inner edge, and hanging out quite expoſed to view: in the lower jaw eight ſmall cutting teeth; and in each jaw ſix grinders: ears long and narrow, inſide of a pale yellow, outſide deep brown: chin yellow: hair on the whole body erect, very long, and each marked with ſhort waves from top to bottom: color near the lower part cinereous, black near the end: the tips ferruginous. The fore part of the neck, in ſome, marked on each ſide with long white ſtripes from the head to the cheſt: back ſtriped with pale brown,

6   reaching

XXI

Tibet . Musk . ─ p. 124 . ─ . N° 63

reaching to the fides: hoofs long, much divided, and black; fpurious hoofs of the fore feet very long: tail an inch long, hid in the hair: the fcrotum of a bright red color; but the penis fo hid as fcarce to be difcovered.

Female lefs than the male: nofe fharper: wants the two tufks, and has two fmall teats.

Inhabits the kingdom of *Tibet*, the province of *Mohang Meng* in *China*, *Tonquin*, and *Bontan*; about the lake *Baikal*, and near the rivers *Jenefea* and *Argun*. Found from *lat.* 60 to 44 or 45; but never wanders fo far fouth, except when forced through hunger, by great falls of fnow, when they migrate to feed on corn and new-grown rice. Inhabit naturally the mountains that are covered with pines, and places the moft wild and difficult of accefs: love folitude: avoid mankind. The chace is a trade of great trouble and danger. If purfued, they feek the higheft fummits, inacceffible to men or dogs.

That noted drug the mufk is produced from the male. It is found in a bag or tumor of the fize of a hen's egg, on the belly of that fex only, kidney-fhaped and pendulous. It is furnifhed with two fmall orifices; the largeft is oblong, the other round; the one is naked, the other covered with long hairs. The mufk is contained in this; for Mr. *Gmelin* tells us, that on fqueezing the tumor, the mufk was forced through the apertures in form of a fat brown matter. The hunters cut off the bag, and tie it up for fale; but are very apt to adulterate the contents, by mixing other matter with it to encreafe the weight. Thefe animals muft be found in great plenty, for *Tavernier* fays, that he bought in one journey 7673 mufk-bags. The mufk of *Tibet* is far fuperior to that of other places, and of courfe much dearer. The flefh of

the

the males is much infected with this drug, but is eaten by the *Ruffians* and *Tartars.* It is ftrongeft in rutting time.

66. BRASILAN.   Cuguacu-ete. *Marcgrave Brafil.* 235.   Cervula furinamenfis, fubrubra albis ma-
                Pifo Brafil. 97.   culis notata. *Seb. Muf.* 1. 71. *tab.*
             Biche de Guiane. *Des Marchais,* iii. 295.   xliv. *Klein quad.* 22. *Briffon quad.* 67.
             Wirrebocerra. *Bancroft Guiana.* 123.

M. about the fize of a roebuck : ears four inches long: the veins very apparent: eye large and black ; noftrils wide : fpace about the mouth black : the hind legs longer than the fore legs: tail fix inches long ; white beneath : hair on the whole body fhort and fmooth : head and neck tawny, mixed with afh-color : back, fides, cheft, and thighs, of a bright ruft-color : lower part of the belly and infide of the thighs white. *Marc-grave* fays, that the throat and under fide of the neck are alfo white. In all other refpects the ftuffed fkin which I examined, agreed with his defcription.

Inhabits *Guiana* and *Brafil* ; are exceffively timid, and moft re-markably active, and fwift ; like goats, they can ftand with all their four legs placed together on the point of a rock. They are frequently feen fwimming the rivers, and at that time are eafily taken. The *Indians* hunt them, and their flefh is efteemed very delicate. The *French* of *Guiana* call them *Biches* or Does, becaufe, notwithftanding their likenefs to deer, both fexes are without horns. M. *de Buffon* accufes *Seba* of an error, in placing this animal in *Surinam* ; but the laft is vindicated by feveral authorities, who have had ocular proof of its exiftence in *Guiana,* &c.

Meminna

XXII

Indian Musk.—p. 127.— N° 67.

Meminna. *Knox hift. Ceylon.* 21. *De Buffon.* xii. 315. Piffay. *Hamilton's voy.*   67. INDIAN.
             E. *Indies*, i. 261.

M. length 1 foot 5; weight 5 lb. $\frac{1}{2}$; of a cinereous olive-color: throat, breaft, and belly white: fides and haunches fpotted, and barred tranfverfely with white: ears large and open: tail very fhort.

Inhabits *Ceylon* and *Java*. A fine drawing of this animal was communicated to me by Mr. *Loten*, late governor in *Ceylon*.

Le Chevrotain des Indes. *De Buffon,*   Tr. indicus. 65. *Klein quad.* 21.   68. GUINEA.
     xii. 315. 341. *tab.* xlii. xliii. *Gmelin,*   Mofchus pygmæus. *Lin. fyft.* 92. LEV.
     Lin. 174.   Mus.
Tragulus Guineenfis. *Briffon quad.* 66.

M. nine inches $\frac{1}{2}$ long: head, legs, and whole upper part of the body, tawny: belly white: no fpurious hoofs: two very broad cutting teeth in the lower jaw: on each fide of them, three others very flender: in the upper jaw two fmall tufks: ears large: tail an inch long.

The fpecimen in the LEVERIAN MUSEUM is ferruginous, mixed with black. The neck and throat ftriped downwards with white.

They are found in the *Eaft Indies,* and feveral of the iflands: in *Java* and *Prince's* ifland. The *Malayes* call them *Kant-chil*; the *Javans, Poet-jang.* The natives catch them in great numbers in little fnares, carry them in cages to market, and fell them for two-pence halfpenny a piece.

The

The horns which *Linnæus* fays are fold as belonging to this animal, are thofe of the *Royal Antelope*, p. 20.

To this genus muft be referred a large fpecies mentioned by *Nieuhoff*, p. 209, found in the ifle of *Formofa*, which he calls ftags, lefs than ours, but without horns.

Arabian Camel.___129. N°69.

No cutting teeth in the upper jaw. Upper lip divided like   IX. CAMEL.
  that of a hare. Six cutting teeth in the lower jaw.
Small hoofs. No fpurious hoofs.

Καμυλος Αραβιος. *Arift. hift. An.* lib. ii.
  c. 1.
Camelus Arabicus. *Plinii lib.* viii. c. 18.
Camel called Hugiun. *Leo Afr.* 338.
Camelus Dromas. *Gefner quad.* 159.
  Pr. *Alp. hift. Ægypt.* i. 223.
Camelus unico in dorfo gibbo, feu Dro-
  medarius. Camel, or Dromedary.
  Raii fyn. quad. 143. *Klein quad.* 42.

Camelus Dromedarius. C. topho dorfi
  unico. *Lin. fyft.* 90.
Le Dromedaire. *De Buffon,* xii. 211.
  tab. ix. *Briffon quad.* 33.
Camel with one bunch. *Pocock's trav.*
  i. 207. *Shaw's trav.* 239. *Ruffel's*
  hift. Aleppo. 56. 57. *Plaifted's jour-*
  nal, 82.
Djammel. *Forfkal,* iv. N°. 12.

69. ARABIAN,
ONE-BUNCHED
DROMEDARY.

C with a fingle bunch on the back: head fmall: ears fhort:
  neck long, flender, and bending: height to the top of the
bunch fix feet fix inches: hair foft: longeft about the neck, under the throat, and about the bunch: color of that on the protuberance dufky: on the other parts a reddifh afh-color: tail
long: the hair on the middle foft: on the fides coarfe, black, and
long: hoofs fmall: feet flat, divided above, but not thorough:
the bottom excessively tough, yet pliant: has fix callofities on
the legs; one on each knee; one on the infide of each foreleg, on
the upper joint; one on the infide of the hind leg, at the bottom
of the thigh; another on the lower part of the breaft: the places
on which the animal refts when it lies down.

The riches of *Arabia,* from the time of *Job* to the prefent. The
patriarch reckoned 6000 camels among his paftoral treafures;

the moderns estimate their wealth by the numbers of these useful animals. Without them great part of *Africa* would be wretched; by them the whole commerce is carried through arid and burning tracts, impassable but by beasts which Providence formed expressly for the scorched deserts. Their soles are adapted to the sands they are to pass over, their toughness and spungy softness preventing them from cracking. Their great powers of sustaining abstinence from drinking, enables them to pass over unwatered tracts for seven or eight days, without requiring the left liquid; *Leo Africanus* says for fifteen. They can discover water by their scent at half a league's distance, and after a long abstinence will hasten towards it, long before their drivers perceive where it lies.

Their patience under hunger is such, that they will travel many days fed only with a few dates, or some small balls of bean or barley-meal; or on the miserable thorny plants they meet with in the deserts.

The largest kind will carry a load of 1000 or 1200 lb. weight. They kneel down to be loaded; but rise the moment they find the burthen equal to their strength: and will not permit an ounce more to be put on. Are most mild and gentle at all times, but when they are in heat: during that period, are seized with a sort of madness, that it is unsafe to approach them: cannot be prevaled on to quicken their pace by blows; but go freely if gently treated; and seem enlivened by the pipe, or any music. In winter they are covered with long hair, which falls off in the spring, and is carefully gathered, being wove into stuffs, and also cloths to cover tents. In summer their hair is short. Before the great heats the owners smear their bodies, to keep off the flies. The

The *Arabs* are very fond of the flesh * of young camels. The milk of these animals is their principal subsistence; and the dung of camels is the fuel used by the Caravans in the travels over the deserts.

This species is common in *Africa*, and the warmer parts of *Asia*; not that it is spread over either of the continents. It is a common beast of burden in *Ægypt*, and along the countries which border on the *Mediterranean* Sea; in the kingdom of *Morocco*, *Sara* or the *Desert*, and in *Æthiopia*: but no where south of those kingdoms. In *Asia* it is equally common, in *Turky* and *Arabia*, but is scarcely seen farther north than *Persia*, being too tender to bear a more severe climate. It is very common in *India*. They are used there for burden as well as carrying men: for the use of the latter, they generally have a pad put on their backs frequently covered with trappings of scarlet cloth, or silk.

PLACE.

There are varieties among the camels. The *Turkman* is the largest and strongest. The *Arabian* is hardy. What is called the *Dromedary*, *Maihary*, and *Raguahl*, is very swift. The common sort travel about thirty miles a day. The last, which has a less bunch, and more delicate shape, and also is much inferior in size, never carries burdens; but is used to ride on. In *Arabia*, they are trained for running-matches: and in many places, for carrying couriers, who can go above one hundred miles a day on them; and that for nine days together†, over burning deserts, unhabitable by any living creature. The *African* camels are the most hardy, having more distant and more dreadful deserts to

* *Athenæus* relates, that the *Persian* monarchs had whole camels served up to their table, *Lib.* iv, *p.* 130. as the *Romans* had whole boars.

† *Leo Afr.* 338.

S 2

pass

pass over than any of the others, from *Numidia* to the kingdom of *Æthiopia*. *She Chin*, a *Chinese* physician, says, that camels are found wild N. W. of his country *.

C. BACTRIAN. TWO-BUNCHED CAMEL.

Καμηλος Βακτρος. *Arist. hist. An.* ii. c. 1.
Camelus Bactrianus. *Plinii lib.* viii. c. 18.
Camel called Becheti. *Leo Afr.* 338.
Camelus. *Gesner quad.* 150. *Pr. Alp. hist. Ægypt.* i. 223. *tab.* xiii.
Camelus duobus in doso tuberibus, seu

Bactrianus. *Raii syn. quad.* 145.
Camelus Bactrianus. C. dorsi tophis duobus. *Lin. syst.* 90. *Klein quad.* 41.
Le Chameau. *De Buffon*, xi. 211. *tab.* xxii. *Brisson quai.* 32.
Persian camel. *Russel's hist. Aleppo*, 57.
Bocht. *Forskal*, iv.

C. with two bunches on the back; in all other respects like the preceding; of which it seems to be a mere variety, and is equally adapted for riding or carrying loads.

WILD. The two-bunched camel is still found wild in the deserts of the temperate parts of *Asia*, particularly in those between *China* and *India*. These are larger and more generous than the domesticated race †.

TAME. This species is extremely hardy, and is very common in *Asia*; and is in great use among the *Tartars* and *Mongols*, as a beast of burden, from the *Caspian* Sea to the Empire of *China*. It bears even in so severe a climate as that of *Siberia*, being found about the lake *Baikal*, where the *Burats* and *Mongols* keep great numbers. They are far less than those which inhabit Western *Tartary*. Here they live during winter on willows and other trees, and are by this diet reduced very lean. They lose their hair in *April*,

* *Du Halde China*, ii. 225.   † *Pallas* Spicil. Zool. fasc. xi. 4. 5.

and

Bactrian Camel. ___ 132.

Llama __ *133* . *N° 70.*

and go naked all *May*, amidſt the froſts of that ſevere climate. To thrive, they muſt have dry ground and ſalt marſhes. Here is a white variety, very ſcarce, and ſacred to the idols and prieſts *.

The *Chineſe* have a ſwift variety, which they call by the expreſſive name of *Fong Kyo Fo*, or *Camels with feet of the wind.* Fat of camels, or, as thoſe people call it, *Oil of Bunches*, being drawn from them, is eſteemed in many diſorders, ſuch as ulcers, numbneſs, and conſumptions †.

This ſpecies of camel is rare in *Arabia*, being an exotic, and only kept by the great men ‡.

Camels have been introduced into *Jamaica* and *Barbadoes*; but, for want of knowlege of their diet and treatment, have in general been of very little ſervice ‖.

Ovis Peruana. *Hernandez An. Mex.* 660. *Marcgrave Braſil.* 243.
Huanucu-Llama. *De Laet*, 328.
Allo-camelus. *Scaligeri.* Ovis Indica. *Geſner quad.* 149.
Llama. *Ovalle Chile. Churchill's coll.* 44, 45. *Ulloa's voy.* i. 478. *Wood's voyage in Dampier's*, iv. 95. *Molina*. 301.

Camelus Glama. C. corpore lævi, topho pectorali. *Lin. ſyſt.* 91.
Camelus Peruvianus *Glama* dictus. *Raii ſyn. quad.* 145.
Le Lama. *De Buffon*, xiii. 16.
Came .s pilis breviſſimis. Le Chameau de Perou. *Briſſon quad.* 34.
Camelus ſpurius. *Klein quad.* 42.

70. LLAMA.

C. with an almoſt even back, ſmall head, fine black eyes, and very long neck, bending very much, with a protube-

* *Pallas*, M.S.
† *Du Halde*, ii. 225.
‡ *Forſkal*, iv. *Niebuhr deſcr. Arabie*, 145.
‖ *Browne's hiſt. Jamaica*, 488. *Ligon's hiſt. Barbadoes*, 58.

rance on the breaſt conſtantly moiſt, with a greaſy exudation, near the junction with the body: in a tame ſtate, with ſmooth ſhort hair; in a wild ſtate, with long coarſe hair *: white, grey, and ruſſet, diſpoſed in ſpots. According to *Hernandez*, yellowiſh, with a black line from the head along the top of the back to the tail, and belly white. The ſpotted may poſſibly be the tame; the laſt, the wild *Llamas*. The tail ſhort: the height from four to four feet and a half: length, from the neck to the tail, ſix feet. The whole animal, according to Mr. *Byron*, weighed 300 lb. In general, the ſhape exactly reſembles a camel, only it wanted the dorſal bunch.

It is the camel of *Peru*; and before the arrival of the *Spaniards*, was the only beaſt of burthen known to the *Indians*. It is as mild, as gentle, and as tractable. We find, that before the introduction of mules †, they were uſed by the *Indians* to plow the land; that at preſent they ſerve to carry burthens of about 100 lb.; that they go with great gravity, and, like their *Spaniſh* maſters, nothing can prevale on them to change their pace. They lie down to be loaden; and when wearied, no blows can provoke them to go on. *Feuillée* ſays, they are ſo capricious, that if ſtruck, they inſtantly ſquat down, and nothing but careſſes can induce them to riſe. When angry, have no other method of revenging injuries than by ſpitting, and they can ejaculate their *ſaliva* to the diſtance of ten paces; if it falls on the ſkin, it raiſes an itching, and a reddiſh ſpot. Their fleſh is eaten, and ſaid to be as good as mutton. The wool has a ſtrong diſagreeable ſcent. They are very ſure-footed; therefore uſed to carry the *Peruvian*

* *Ulloa*, i. 479.   † *Ovalle*, 44.

4   ores

ores over the ruggedeft hills and narroweft paths of the *Andes*. They inhabit that vaft chain of mountains, their whole length, to the ftraits of *Magellan*; but, except where thofe hills approach the fea, as in *Patagonia*, never appear on the coafts. Like the camel, they have powers of abftaining long from drink, fometimes for four or five days: like that animal's, their food is coarfe and trifling.

Molina, who had frequent opportunity of feeing thofe animals in their native country, affures us that they differ fpecifically from the *Guanaco*. *Linnæus* had united them, but we muft give way to the evidence of eye-witneffes.

This and every other fpecies of *South America* inhabit the fnowy *Andes* and *Cordillera*. Their bodies are covered with fat between the fkin and the flefh: and they abound with blood: both requifite to preferve warmth in their frozen refidence.

They keep in great herds, in the higheft and fteepeft parts of the hills, and alfo near the fhores; and while they are feeding, one keeps centry on the pinnacle of fome rock: if it perceives the approach of any one, it neighs; the herd takes the alarm, and goes off with incredible fpeed. When they get to a confiderable diftance will ftop, look at their purfuers till they come near, and then fet off again *. They out-run all dogs; fo there is no other way of killing them but with a gun. They are killed for the fake of their flefh and their hair; for the *Indians* weave the laft into cloth †. From the form of the parts of generation, in both fexes, no animal copulates with fuch difficulty: it is often the labor of a day, *Antequam actum ipfum venereum incipiant, et abfolvant ‡.

* *Byron's voy.* 18.
† *De Laet*, 329. ‡ *Hernandez*, 662.

L 2

71. VICUNNA.

La vigogne *Molina*, 295. *Schreber, tab.* cccvii.
Ovis Chilenfis. *Wood's voy. Dampier,* iv. 95. *Narborough's voy.* 32.
Vicunna, Alpaques. *Frezier's voy.* 153, 154. *Ulloa's voy.* i. 479.
Camelus feu Camelo congener Peruvianum lanigerum, *Pacos* dictum. *Raii fyn.*

quad. 147.
Camelus Laniger. *Klein quad.* 42.
Le Paco. *De Buffon,* xiii. 16.
Camelus pilis prolixis toto corpore veftitus. Le vigogne. *Briffon quad.* 35.
Camelus Pacos. :. tophis nullis, corpore lanato. *Lin. fyft.* 91.

C. with the body covered with long and very fine wool, of the color of dried rofes, or a dull purple: the belly white. Head round, nofe fhort, tail like that of a goat. In a tame ftate, varies in color. Shaped like the former, but much lefs: the leg of one I faw was about the fize of that of a buck.

Are of the fame nature with the preceding: inhabit the fame places, but are more capable of fupporting the rigor of froft and fnow: they live in vaft herds; are very timid, and exceffively fwift: fometimes the *Guanacoes* affociate with them. The wool is very valuable both in *Chili* and in *Europe*, and is fufceptible of any dye. The flefh is excellent eating. The *Indians* take the *Pacos* in a ftrange manner: they tie cords, with bits of wool or cloth hanging to them, above three or four feet from the ground, crofs the narrow paffes of the mountains, then drive thofe animals towards them, which are fo terrified by the flutter of the rags as not to dare to pafs, but huddling together, give the hunters opportunity to kill with their flings as many as they pleafe. Thefe animals are not yet domefticated.

Thefe animals yield a *Bezoar: Wafer* * fays he has taken thir-

* *Wafer's voy.* in *Dampier,* iii. 384.

The Vicunna __ 136. __ N.º 71.

teen out of the ftomach of a fingle breaſt: they were ragged, and of feveral forms, fome round, fome oval, others long: they were green at firſt, but changed to aſh-color.

Le Paco ou Alpaco. *Molina, 296.* Camelus Paco. *Gmelin, Lin.* 171.   72. PACOS.

C. with an oblong vifage: body covered with very long wool: of a make more robuſt than the *vicunna.*

Inhabits *Peru* only; the natives keep vaſt flocks of them for the fake of the wool, which they work into ſtuffs as refplendent as filks. They ſerve alfo to carry burdens; and, like camels, they bend their knees to receive or difcharge their loads. They are found on the mountains of *Peru* in a ſtate of nature, as well as the *vicunna,* but never mix together. This deſtroys the opinion M. *de Buffon* had, that the *paco* and *vicunna* were the fame animal, and that the firſt was only a wild *vicunna.* Father *Molina* fatisfies us of that miſtake: he befides adds three more of *American* camels to the two we were before acquainted with. That gentleman was a jefuit, refident in *South America,* who had formed great collections in Natural Hiſtory. When the order was expelled out of the new world, the *Spaniards* deprived him of every thing. By a ſtrange accident on his return (I think to *Bologna,* his native place) he recovered one of his manufcripts, which was tranſlated out of the *Italian* into *French* under the title of *Effai ſur l'Hiſtorie Naturelle du Chili,* and publiſhed at *Paris* in 1789, in octavo. It is a choice and inſtructive work; which gives us great reaſon to regret the loſs of the reſt of his labors.

VOL. I.   T   Camelus

73. GUANACO.   Camelus Huanacus. *Molina*, 300. *Gmelin, Lin.* 170. *Schreber*, tab. cccvi.

C. with a round head, pointed nofe: long hair; tawny on the back, white on the belly: back arched: tail fhort, and turned upwards: ears ftrait like thofe of a horfe: the hind legs very long: fometimes grows to the fize of a horfe.

Inhabits, during fummer, the tops of the mountains; but more tender than the *Pacos:* defcends in winter into the vallies. It runs with amazing fwiftnefs; and, from the great length of the hind legs, prefers defcending the hills, which it does by leaps and bounds like the buck. When young it is hunted and taken with dogs; when old, they are chaced by the *Indians* mounted on fwift horfes, who catch them with noofes, which they fling with great dexterity. Thefe animals are eafily domefticated: their flefh is excellent when young: in an adult ftate it is falted, and is capable of very long prefervation.

74. CHILI-   *Molina*, 298. Camelus araucanus. *Gmel. Lin.* 1702.
HUCQUE.

C. with a head like a fheep, ears oval, and lips thick and pendulous; nofe long, and arched: tail like that of a fheep: body covered with long wool, very foft: length fix feet; height four: varies in color (I fuppofe in a domeftic ftate) to white, brown, black, and grey.

Thefe animals inhabit *Chili*, and were employed by the antient *Chilians* as beafts of burden. They were led by a cord paffed

6

through

through the nose. Before the conqueſt of *America* the wool was manufactured in cloth, but is difuſed ſince the introduction of ſheep. The *Chilians* love the fleſh, but never kill the animal but on great feaſts or ſolemn ſacrifices.

This is the *ovis chilenſis* of *Ovalle.* p. 44. *Cieza,* 232. and *Feuillé,* iii. 23. and *Marcgrave,* 244.

Cutting

X. HOG.   Cutting teeth in both jaws.

75. COMMON. *(Wild).* Sus fera, aper. *Plinii lib.* viii.   *(Tame).* Sus. *Gesner quad.* 872. *Raii syn.*
c. 51. *Gesner quad.* 918.   *quad.* 92.
Sus agrestis sive aper, wild boar or swine.   Schwein. *Klein quad.* 25.
Raii syn. quad. 96.   Le Cochon. *De Buffon*, v. 99. Le Ver-
Wieprz lesny, Dzik. *Rzaczynski Polon.*   rat. *tab.* xvi.
213.   Sus caudatus, auriculis oblongis, acutis,
Wild Schwein. *Klein quad.* 25.   cauda pilosa. *Brisson quad.* 74.
Le Sanglier. *De Buffon*, v. 99. *tab.* xiv.   Sus scrofa. S. dorso antice setoso, cauda
Sus caudatus, auriculis brevibus, subro-   pilosa. *Lin. syst.* 102. Swiin. *Faun.*
tundis, cauda pilosa. *Brisson quad.*   *suec.* N° 21. *Br. Zool.* i. 41. LEV.
75.   MUS.
Sus aper. *Lin. syst.* 102.

WILD.   H with the body covered with bristles : two large tusks
              above and below : six cutting teeth in each jaw. In
a wild state, of a dark brinded color : beneath the bristles is a
soft curled short hair : the ears short, and a little rounded.
TAME.   TAME : the ears long, sharp-pointed, and slouching : the color
generally white, sometimes mixed with other colors.
  The *Siam* Hog of M. *de Buffon* is a variety, differing chiefly in
the superior length of the tail.
PLACE.   In a tame state, universal, except in the frigid zones, and
*Kamtschatka**, and such places where the cold is very severe.
Since its introduction into *America*, by the *Europeans*, abounds to
excess in the hot and temperate parts. Found wild in most parts
of *Europe*, except the *British* isles, and the countries N. of the

* *Hist. Kamtf.* 108.

Baltic :

Variety of Common Hog — 140 . N°75

Baltic: in *Asia*, from *Syria* to the borders of the lake *Baikal*[*], and as high as 55° N. latitude: in *Africa*, on the coast of *Barbary*. Are very numerous in *Ceylon*, *Celebes*, and *Java*; but are generally leſſer than the *European*, yet are of the same ſpecies. In the foreſts of *South America* [†] are vaſt droves, which derive their origin from the *European* kind relapſed into a ſtate of nature, and are what Mr. *Bancroft*, in his hiſtory of *Guiana*, 126, deſcribes as a particular ſpecies, by the name of *Warree*. Inhabits wooded countries: very ſwift: a ſtupid, ſlothful, drowſy animal, fond of wallowing in the mud to cool its ſurfeited body. Greedy, voracious, but not indiſcriminate in the choice of its food; has been found to eat 72 ſpecies of plants, reject 171: very fond of various roots: ſo brutal as to eat its own offspring. Uſeful in *America*, by clearing the country of rattle-ſnakes, which it devours with ſafety. Reſtleſs in high winds: has a natural diſpoſition to grow fat: is very prolific, brings ſometimes 20 young at a time. Its fleſh of vaſt uſe; takes ſalt the beſt of any; furniſhes our table with various delicacies; *brawn*, peculiar to the *Engliſh*. The *Romans* made a diſh..

Of the ſwelling unctuous paps
Of a fat pregnant Sow, newly cut off[‡].

[*] *Bell's trav.* i. 279.
[†] *Des Marchais voy.* iii. 312. *Gumilla Orenoque*, ii. 4.
[‡] *Alchymiſt, Act.* ii. Sc. ii.

a GUINEA:

α GUINEA. Porcus Guineensis. *Marc-* | longitudine pedum. *Lin. syst.* 103.
grave Brasil. 230. *Raii syn. quad.* 96. | Le Cochon de Guinéa. *De Buffon,* xv.
Sus porcus. S. dorso postice setoso, cauda | 146. *Brisson. quad.* 76.

H. with a lesser head than the common kind: very long, slender, and sharp-pointed ears: tail hanging down to the heels, without hairs: the body covered with short, red, shining hairs, but about the neck and lower part of the back a little longer: no bristles. A domestic variety of the common kind.

β The *Siam* hog is another variety, very little differing from the former. It is described by *M. de Buffon,* under the title of *Cochon de Siam.* v. 99. tab. xv.

γ CHINESE. Sus Chinensis. *Lin. syst.* 102. *Brisson, quad.* 75. Javan Hog. *Kolben Cape* i. 117.

H. with the belly hanging almost to the ground: legs short: tail very short: the body generally bare, as is the case in general with the swine of *India.*

Its wild breed is found in great numbers in *New Guinea,* and in the islands of that country, which the *Papuas* chace in their canoes, as the animals are swimming from island to island, and kill them with lances, or shoot them with arrows *. They are also found on the island of *Gilolo,* and resort eagerly to the places where *sago* trees have lately been cut down, to feed on the pith

* *Forrest's* Voy. tab. xi. and p. 97.

left

left there, which makes them very fat. They are said to appear, with their little black pigs, like so many flies on a table*

New Guinea must originally have supplied with hogs such of the islands of the *South Sea*, which are happy enough to possess these animals. They passed first to the *New Hebrides*, thence to the *Friendly Isles*, the *Society*, and the *Marquesas*. All the islands to the east, and even *New Caledonia*, a little to the south, are destitute of them. They are of the same variety with the *Chinese*, and are more delicious food, being fed with plantanes, bread fruit, and yams: but are often too fat for an *European* stomach.

They are the animals which are sacrificed to the lesser deities of the isles: are roasted whole, placed on altars, and left there to decay.

The priests support my notion of the place of their origin: men, dogs, hogs, poultry, and rats, say they, came originally from an island, which they style the *Mother of Lands*: i. e. some island comparatively vastly larger than their own. This island is probably *New Guinea*, where the same species of hog, and the currish fox-like dog, are found. As Captain *Forrest* informed me that *New Guinea* is not destitute of rats, it is not unlikely but that they were imported by some of the early navigators, and, escaping from the ships, became the pest of the islands.

* *Forrest's* Voy. p. 39.

᷂. H.

δ. H. with undivided hoofs, only a variety of the common kind.

76. ÆTHIOPIAN.

Engalla. *Sorrento's voy. in Churchill,* i. 667. *Barbot.* 487. *Dampier's voy.?* i. 320.
African wild boar. *Deflande's Martyn's mem. Acad.* v. 386.
Sus Æthiopicus, Hardlooper. *Pallas m'fcel. zool.* 16. *tab.* xi. *fpicil. fafc.* .ii.

1. *tab.* i. *Flacourt hiſt. Madagaſcar.* 511.
Sus Æthiopicus. S. facculo molli fub oculis. *Lin. ſyſt. App. tom.* III. 223.
Sanglier du Cap. Verd. *de Buffon.* fupplem. iii. 76. *tab.* xi. *Journal Hiſtorique,* tab. p. 62. LEV. MUS.

H. with fmall tufks in the lower jaws; very large ones in the upper; in old boars bending up towards the forehead, in form of a femicircle. As a fingular mark of this fpecies, it has no fore teeth, their place being occupied by very hard gums.

The nofe is broad, depreffed, and almoft of a horny hardnefs: head very large and broad: beneath each eye a hollow, formed of loofe fkin, very foft, and wrinkled; under thefe a great lobe or wattle, lying almoft horizontal, broad, flat, and rounded at the end, placed fo as to intercept the view of any thing below from the animal.

Between thefe and the mouth, on each fide, is a hard callous protuberance: mouth fmall: fkin dufky: briftles difpofed in *fafciculi,* of about five each; longeft between the ears, and on the beginning of the back, and but thinly difperfed on the reft of the back.

Ears large and fharp-pointed, infide lined with long whitifh hairs: tail flender and flat; does not reach lower than the thighs, and covered with hairs difpofed in *fafciculi.*

Body

Body longer, and legs shorter, than in the common swine: its whole length four feet nine inches; height before two feet two.

These animals inhabit the hottest parts of *Africa*, from *Sierra Leone* to *Congo*, and to within about two hundred leagues of the Cape. The *Hottentots* call them *Kaunoba*. They are found also in the island of *Madagascar* *.

We also suspect that they are found in the isle of *Mindanao*, for *Dampier* † says that the hogs of that island are very ugly creatures, with great knobs growing over their eyes: that there are multitudes of them in the woods, and that they are commonly very poor, but sweet.

It lives under ground ‡; and burrows as expeditiously as the mole, forming almost instantaneously a great hole in the ground, by means of its callous snout, as was experienced from the animal preserved in the Prince of *Orange*'s menagery at the *Hague*.

We know little of their manners; but they are represented as very fierce and swift; and that they will not breed either with the domestic or *Chinese* sow; for that at the *Hague* killed one of the last, and treated the other very roughly, which for experiment were turned to it ||. Its savage nature proved fatal to its keeper, whom it slew, by a wound in the thigh.

The *Hottentots* dread the attack of them more than that of the lion. If not timely repelled, they will rush on a man, snap his legs in two, or rip open his belly: when the old ones are closely pursued, with their young, each will catch up a pig in its mouth, and convey it to a place of security.

* Ces sangliers, principalement les masles, ont deux cornes a costez de nez qui sont comme deux callositez. *Flacourt hist. Madag.* 152.

† *Voy.* i. 321. ‡ *Sparman.* || *Vosmaer Monogr.*

Sanglier

77. CAPE VERD.   Sanglier de Cape Verd. *De Buffon*, xiv. 409. xv. 148. *Aſh. Muſ.* (the jaws only.)
LEV. MUS.

H. with two cutting teeth in the upper, and ſix in the lower jaw. Six grinding teeth on each ſide in both: the fartheſt very large: twenty-four in all. The tuſks large, and of the hardneſs of ivory. The tuſks of the upper jaw thick, and truncated obliquely.

Head long, noſe ſlender: upper jaw extends far beyond the lower. Ears narrow, upright, pointed, and tufted with very long briſtles. The whole body covered with very long fine briſtles, eſpecially about the ſhoulders, belly, and thighs, where they are of great length. The tail ſlender, and terminating in a large tuft. It reaches to the firſt joint of the leg.

Inhabits *Africa*, from *Cape Verd* to that of *Good Hope*. Seems to be the ſame with that ſeen by Mr. *Adanſon*, who calls it a boar of enormous ſize, peculiar to *Africa*.

I believe that the only entire ſpecimen of the head now in *Europe*, is in poſſeſſion of Sir *Aſhton Lever*, which he received from the *Cape*.

Quauhtlá.

Quauhtla. coymatl. Quapizotl. aper
  Mexicanus. *Hernandez an. mex.* 637.
Hogs with navels on their backs. *Pur-*
  chas's Pilgr. iii. 868. 966.
Tajacu. *Piſo Braſil.* 98. *Barrere France*
  equin. 161.
Tajacu, Caaigora. *Marcgrave Braſil.* 229.
  Ovalle Chile, Churchill, iii. 2.
Tajàcu ſeu aper Mexicanus moſchiferus.
  Raii ſyn. quad. 77.
Mexican muſk hog. *Ph. Tr. abr,* ii. 876.

Pecary. *Wafer's voy. Dampier.* iii. 328.
  iv. 48. *Rogers's voy.* 345.
Des Marchais voy. iii. 312. *Gumilla Ore-*
  noque, ii. 6. *Bancroft Guiana,* 124. *De*
  Buffon, x. 21. *tab.* iii. iv. *Seb. muſ.* i.
  177.
Javaris. *Rochfort Antilles.* i. 285.
Sus ecaudatus, folliculum ichoroſum in
  dorſo gerens. *Briſſon quad.* 77.
Sus dorſo cyſtifero, cauda nulla.
S. Tajacu. *Lin. ſyſt.* 103. LEV. MUS.

78. MEXICAN.

H. with four cutting teeth above, ſix below; two tuſks in
each jaw; thoſe in the upper jaw pointing down, and
little apparent when the mouth is ſhut; the others hid. Length
from noſe to the end of the rump about three feet: head not ſo
taper as in common ſwine: ears ſhort and erect: body covered
with briſtles, ſtronger than thoſe of the *European* kind, and more
like thoſe of a hedge-hog; they are duſky, ſurrounded with rings
of white; thoſe on the top of the neck and back are near five
inches long, grow ſhorter on the ſides: the belly almoſt naked:
from the ſhoulders to the breaſt is a band of white: no tail: on
the lower part of the back is a gland, open at the top, diſcharg-
ing a fœtid ichorous liquor; this has been miſtakenly called a
navel.

Inhabits the hotteſt parts of *S. America,* and ſome of the *An-*
tilles: lives in the foreſts on the mountains: not fond of mire or
marſhy places: leſs fat than the common hog: goes in great
droves: is very fierce: will fight ſtoutly with the beaſts of
prey: the *Jaguar,* or *American* leopard, is its mortal enemy;

often

often the body of that animal is found surrounded with thofe of numbers of thefe hogs, all flain in combat. Dogs will fcarcely attack it : if wounded, it will turn on the hunter. Feeds on fruits and roots, on toads, and all manner of ferpents, and holding them with the fore-feet, fkins them with great dexterity. Is reckoned very good food; but all writers agree that the dorfal gland muft be cut out as foon as the animal is killed, or the flefh will become fo infected as not to be eatable. The *Indian* name of this fpecies is *Paquiras* *, from whence feems to be derived that of *Pecary*.

79. BABY-ROUSSA.

Aper in *India*, &c. *Plinii lib.* viii. *c.* 52. Υς τετράκερως. *Ælian. an. lib.* xviii. *c.* 10. Baby-rouffa. *Bontius India.* 61, *Grew's Mufeum.* 27. *Raii fyn. quad.* 96. *Klein quad.* 25. *Seb. Muf.* i. 80. *tab.* l. *Valentyn Amboin.* iii. 268.
Strange hog. *Purchas's Pilgr.* ii. 1693. v. 566. *Nieuhoff's voy.* 195.

Sus dentibus duobus caninis fronti innatis. S. Babyruffa. *Lin. fyft.* 104. Sus caudatus, dentibus caninis fuperioribus, ab origine furfum verfis, arcuatis, cauda floccofa. *Briffon quad.* 76.
Le Babirouffa. *De Buffon.* xii. 379. *tab.* xlviii. *Br. muf. Afhm. muf.* LEV. MUS.

H. with four cutting teeth in the upper, fix in the lower jaw; ten grinders to each jaw; in the lower jaw two tufks pointing towards the eyes, and ftanding near eight inches out of their fockets; from two fockets on the outfide of the upper jaw, two other teeth, twelve inches long, bending like horns, their ends almoft touching the forehead : ears fmall, erect, fharppointed : along the back are fome weak briftles : on the reft of the body only a fort of wool, fuch as is on lambs : the tail long,

* *Gumilla.*

ends

Babyroussa —— 148. — S.º 79.

ends in a tuft, and is often twifted : the body plump and fquare; not of the elegant form which *Bontius* and *Nieuhoff* give it; as appears by an original drawing Mr. *Loten* favored me with.

Inhabits *Boero*, a fmall ifle near *Amboina*: but neither on the continent of *Afia*, or *Africa*; what M. *de Buffon* takes for it, is the *Æthiopian* boar. They are fometimes kept tame in the *Indian* ifles: live in herds: have a very quick fcent: feed on herbs and leaves of trees; never ravage gardens, like other fwine : their flefh well-tafted. When purfued, and driven to extremities, rufh into the fea, fwim very well, and even dive, and pafs thus from ifle to ifle : in the forefts often reft their head, by hooking their upper tufks on fome bough *. The tufks, from their form, ufelefs in fight.

* The natural hiftory of this animal is taken from *Valentine's hift.* of the *Eaft Indies*, from a tranflation Mr. *Loten* was fo obliging to communicate to me.

With

With one, fometimes two, large horns on the nofe.
Each hoof cloven into three parts.

80. TWO-HORNED.   Rhinoceros cornu gemino. *Martial fpec-*   p. 103.
tac. ep. 22.   Ph. Tr. Abr. ix. 100.   *Flacourt, hift. Madag.* 395. *De Buffon.*
xi. 910. *Ph. Tr. vol.* lvi. 32. *tab.* ii.   xi. 186. *Lobo, Abyfi.* 230.
Kolben, ii. 101.   Rhinoceros bicornis. *Lin. fyft.* 104. *Br.*
Sparman, Stock. wettfk. Handl. 1778.   *muf.* LEV. MUS.

R H. with two horns, one placed beyond the other. Length of
the fore horn of one in the Ph. Tranf. twenty inches, of the
fecond horn nineteen; but they vary in fizes. Upper lip fhort,
reaching but a little way over the lower: no fore teeth. The fkin
without any *plicæ* or folds; much granulated or warty; of a deep
cinereous grey. Between the legs fmooth, and flefh-colored. In
other parts are a few fcattered ftiff briftles, moft numerous about
the ears and end of the tail. Tail thick as a thumb: convex
above and below: flatted on the fides. Feet no more in diameter
than the legs: but the three hoofs project forward. Soles callous.

PLACE.
Inhabits *Africa.* Obferved firft by *Flacourt,* in the bay of
Saldagne, near the Cape. Within thefe few years by Mr. *Sparman,* a learned *Swede,* at fome diftance N. of that promontory.
He, with the laudable perfeverance of a naturalift, watched the
arrival of thofe and other animals at a muddy water, whither the
wild beafts refort to quench their thirft, and fome to indulge, in
that hot climate, in rolling in the mud. In that fpot he fhot two
of thefe animals: one was fo large that the united force of five
men could not turn it. The leffer he meafured: its length
                                                           was

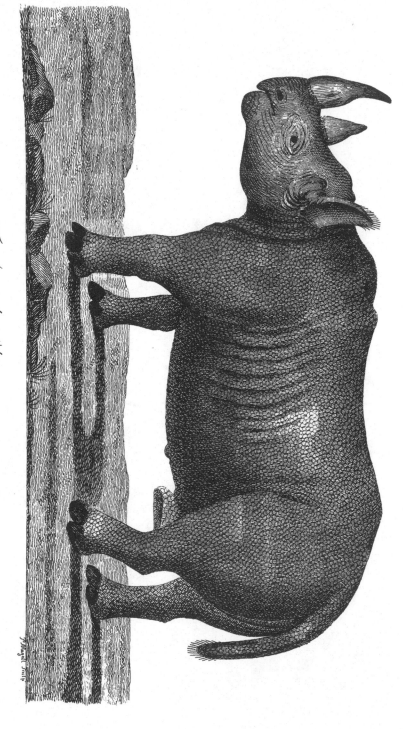

Two horned Rhinoceros — 150. 1.80

was eleven feet and a half, the girth twelve: the height, between six and seven.

The skin is quite naked, very strong and thick, but is easily penetrated with an iron bullet: one of lead is flatten'd against the hide. The *Hottentots* at prefent always kill thefe animals by a mufquet shot, and the skin is capable of being transfixed with the launce or dart. The *Hottentots* usually hasten the death of the *Rhinoceros*, by taking care to poifon the weapon.

This species feems to agree in manners with the following. Its flesh is eatable, and tastes like coarfe pork. Cups are made of the horns; and of the hide, whips. Its food is boughs of trees, which it bites into bits of the fize of a finger. It feeds alfo much on fucculent plants, efpecially the ftinking *ftapelia*, and a fpecies of *Stæbe* called the *Stæbe Rhinocerotis*.

It continues during day in a state of reft. In the evenings and mornings (perhaps the whole night) wanders in queft of food: or in fearch of places to roll in.

Has no voice, only a fort of fnorting, which was obferved in females anxious for their young.

Its dung is like that of horfes. It has a great propenfity to cleanlinefs, dropping its dung and urine only in particular places.

Its fenfe of fight is bad. Thofe of hearing and fmelling very exquifite: the left noife or fcent puts it in motion. It inftantly runs to the fpot from which thofe two fenfes take the alarm. Whatfoever it meets with in its courfe, it overturns and tramples on. Men, oxen, and waggons, have thus been overturned, and fometimes deftroyed. It never returns to repeat the charge; but keeps on its way: fo that a fenfelefs impulfe, more than rage, feems the caufe of the mifchief it does.

This

RHINOCEROS.

This was the species described by *Martial,* under the name of RHINOCEROS *cornu gemino*: who relates its combat with the Bear.

Namque gravem gemino cornu fic extulit urfum,
Jaɔtat ut impofitas taurus in aftra pilas *.

In fact, the *Romans* procured their *Rhinoceroses* from *Africa* only, which was the reason why they are reprefented with double horns. That figured in the *Preneftine* pavement, and that on a coin of *Domitian,* have two horns: that which *Paufanias†* defcribes under the name of *Æthiopian* Bull had one horn on the nofe, and another leffer higher up: and *Cofmas Ægyptius‡,* who travelled into *Æthiopia,* in the reign of *Juftinian,* alfo attributes to it the fame number: whereas *Pliny,* who defcribes the *Indian* kind, juftly gives it but a fingle horn. *Cofmas,* vol. II. p. 334, fays, that its fkin was fo thick and hard, that the *Æthiopians* ploughed with it, and that they called the animal *Aru* and *Harifi*: the laft fignifying the figure of the noftrils, and the ufe made of the fkin. He adds, that when the beaft is quiefcent, the horns are loofe, but in its rage become firm and immoveable. This is confirmed by Doɔtor *Sparman,* who obferved that they were fixed to the head, or rather nofe, by a ftrong apparatus of finews and mufcles, fo as to give the animal the power of giving a fteady fixture whenever occafion demands.

Auguftus introduced a *rhinoceros* (probably of this kind) into the fhews, on occafion of his triumph over *Cleopatra* ‖.

---

* Speɔt. Epig. 22.   † ix. 9.   ‡ Tom. ii. 334.   ‖ *Dion Caffius,* lib. li.

Mr.

Mr. *Bruce's* figure of a *Rhinoceros* lies * under fome fufpicion of being moft faithfully copied from the fingle horned fpecies of M. *de Buffon*†, with the long upper lip and every characteriftic fold and plait: but by the addition of another horn, it becomes *Bicornis*; and, as Mr. *Bruce* very juftly twice obferves, the firft drawing of the kind ever prefented to the public ‡. So true is the old faying, *Semper aliquid novi* AFRICAM *afferre!*

I am indebted to Mr. *Paterfon* for my figure of the two-horned fpecies: it does not differ materially from that by Doctor *Sparman*, unlefs in the lateral marks that diftinguifh the former: and feem no more than a loofenefs of fkin. M. *Allamand* had engraved the fame animal from a drawing communicated to him by Col. *Gordon*, the great explorer of *Caffraria*; and M. *de Buffon* again copied his plate from a drawing, in which the loofenefs of the fkin on the fides is far better expreffed ‖.

I will not quit the fubject till I have laid before the public my reafons to imagine that this fpecies is not confined to *Africa*. Mr. *William Hudfon*, with his ufual friendfhip, communicated to me the following remark of Mr. *Charles Miller*, who was long refident in *Sumatra*: ' I never faw but two of the two-horned ' *Rhinoceros*; but I believe they are not uncommon in the ifland, ' but are very fhy, which is the reafon they are but feldom ' feen. I was once within twenty yards of one. It had not ' any appearance of folds or plaits on the fkin; and had a fmaller ' horn refembling the greater, and, like that, a little turned in- ' ward. The figure given by Doctor *Sparman* is a faithful re- ' femblance of that I faw.'

---

* Vol. v. tab. p. 85.   † Vol. xi. tab. vii.
‡ Vol. v. p. 86. 87.   ‖ *De Buffon Supplem.* vi. 78. tab. vi.

81. ONE-HORNED.

Rhinoceros. *Plinii lib.* viii. *c.* 20. *Geſ-ner quad.* 842. *Raii ſyn. quad.* 122. *Klein quad.* 26. *Grew's muſeum,* 29. *Worm. muſ.* 336. *De Buffon,* xi. 174. *tab.* vii. *Briſſon quad.* 78. *Ph. Tr. Abr.* ix. 93. *Schreber,* ii. 44. tab. lxxviii.

Rhinoceros or Abbados. *Linſchotten Itin.* 56. *Bontius India.* 50. *Eorri biſt. Cochin-Chinæ.* 797. *Du Halde China.* i. 120. *Faunul. Sinens.*

Rhinoceros unicornis. *Lin. ſyſt.* 104. *Edw.* 221. *Br. Muſ. Aſh. Muſ. Lev. Mus.*

RH. with a ſingle horn, placed near the end of the noſe, ſome-times three feet and a half long, black and ſmooth: the upper lip long, hangs over the lower, ends in a point; is very pliable, and ſerves to collect its food, and deliver it into the mouth: the noſtrils placed tranſverſely: four cutting teeth; one on each corner of each jaw. Six grinders in each; the firſt re-mote from the cutting teeth. The ears large, erect, pointed: eyes ſmall and dull: the ſkin naked, rough, or tuberculated, thick and ſtrong, lying about the neck in vaſt folds; there is another fold from the ſhoulders to the fore-legs; another from the hind part of the back to the thighs: the tail is ſlender, flatted at the end, and covered on the ſides with very ſtiff thick black hairs: the belly hangs low: the legs ſhort, ſtrong, and thick: the hoofs divided into three parts; each pointing forward.

Thoſe which have been brought to *Europe* have been young and ſmall: *Bontius* ſays, that in reſpect to bulk of body, they equal the elephant, but are lower on account of the ſhortneſs of the legs.

Inhabits *Bengal, Siam, Cochin-China, Quangſi* in *China,* and the iſles of *Java* and *Sumatra;* loves ſhady foreſts, the neighborhood of rivers, and marſhy places: fond of wallowing in mire, like

the

the hog; is faid by that means to give fhelter in the folds of its fkin to fcorpions, centipes, and other infects. Is a folitary animal: brings one young at a time, very folicitous about it: quiet and inoffenfive; but when provoked, furious: very fwift, and very dangerous: I know a gentleman * who had his belly ripped up by one, but furvived the wound. Is dull of fight; but has a moft exquifite fcent: feeds on vegetables, particularly fhrubs, broom, and thiftles: grunts like a hog: is faid to confort with the tiger; a fable, founded on their common attachment to the fides of rivers, and on that account are fometimes found near each other.

It is faid, when it has flung down a man, to lick the flefh quite from the bone with its tongue: this is impoffible, as the tongue is quite fmooth; that which wounded the gentleman, retired in-ftantly after the ftroke.

Its flefh is eaten; the fkin, the flefh, hoofs, teeth, and very dung, ufed in *India* medicinally; the horn is in great repute as an antidote againft poifon †, efpecially that of a virgin *Abbada*; cups are made of them, which are fuppofed to communicate the virtue to the liquor poured into them.

THE UNICORN.

Is the unicorn of HOLY WRIT, and *Indian* afs of *Ariftotle* ‡, who fays, it has but one horn; his informers might well compare the clumfy fhape of the *Rhinoceros* to that of an afs, fo that the philofopher might eafily be induced to pronounce it a whole-footed animal. I may add, that *Ælian, lib.* iv. *c.* 22, attributes the fame *alexipharmic* qualities to the horn of the *Indian* afs, as

---

* *Charles Pigot*, Efq; of *Peploe*, *Shropfhire*, at that time in the *India* fervice.

† It was not every horn that had this virtue: fome were held very cheap, while others take a vaft price.

‡ *Hift. An. lib.* ii. *c.* 1.

X 2

are

are afcribed to that of the *Rhinoceros*. This was alfo the *fera monoceros* of *Pliny* *; which was of *India*, the fame country with this animal; and in his account of the *monoceros*, he exactly defcribes the great black horn and the hog-like tail. The *unicorn* of HOLY WRIT has all the properties of the *Rhinoceros*, rage, untameablenefs, great fwiftnefs, and great ftrength.

Various animals were ftyled *monoceros* and *unicornis*, probably from the accident of having loft one of their horns. Thus *Pliny* mentions a *bos unicornis*, and *oryx unicorne*. Any of the great ftrait-horned antelopes, fuch as the *Indian*, N° 22, deprived of one horn, would make an excellent *unicorn*, and anfwer to the figure given of it: for on fuch an accident the fable feems to be founded, when the word is not applied to the *Rhinoceros*.

The combats between the Elephant and Rhinoceros, a fable, derived from *Pliny*.

An entire *Rhinoceros* was found buried in a bank of a *Siberian* river, in the antient frozen foil, with the fkin, tendons, and fome of the flefh in the higheft prefervation. This fact, incredible as it is at firft fight, is given, not only on the beft authority †: but, as an evidence, the complete head is now preferved in the *Mufeum* at *Peterfburg:* the body was difcovered in 1772, in the fandy banks of the *Witim*, a river falling into the *Lena* below *Jakutfk*, in N. lat. 64, and a moft ample account of it given by that able naturalift Doctor PALLAS, to whom this work is under fuch frequent obligations.

* *Lib.* viii. *c.* 21.   † Dr PALLAS, Nov. Com. *Petrop.* xvii. 585. tab. xv.

Four

Male Hippopotame — 157. N.º 82.

Four cutting teeth in each jaw : two tusks in each. Each hoof divided into four parts.

Ἵππος ποταμιος, *Ariſtot. hiſt. An. lib.* ii. *c.* 7.

Hippopotamus. *Plinii lib.* viii. *c.* 26. *Belon obſ.* 104. *des Poiſſons*, 19, 20. *Geſner quad.* 493. *Radzivil iter Hieroſol.* 142. *Rai ſyn. quad.* 123.

River Horſe, or Hippopotamus. *Grew's Muſeum*, 14. *tab.* 1. *Ludolph. Æthiop.* 60.

Cheropotamus et Hippopotamus. *Proſp. Alp. hiſt. Ægypt.* i. 245.

Sea Horſe. *Leo Afr.* 344. Sea Ox. *ibid. Lobo Abyſſ.* 105. *Kolben Cape*, ii. 129.

Hippopotamus, or Behemoth. *Shaw's trav. Suppl.* 87.

T'gao of the *Hottentots.*

Sea Horſe. *Dampier's voy.* ii. 104. *Adanſon's voy.* 133. *Moore's voy. Gambia*, 105, 188, 216.

River-Paard. *Houttuyn, Nat. hiſt.* iii. 405. *tab.* xxviii.

Water Elephants. *Barbot voy. Guinea*, 113, 73.

Hippopotamus pedibus quadrilobis. H. amphibius. *Lin. ſyſt.* 101. *Haſſelquiſt iter*, 201. *Klein quad.* 34. *Journal hiſtorique, &c.* 17. *tab.* ii. *Allamande*, 124.

L'Hippopotame. *De Buffon*, xii. 22. *tab.* cxi. *Briſſon quad.* 83. *Br. Muſ. Aſhm. Muſ.* LEV. MUS.

H. with four cutting teeth in each jaw ; thoſe in the lower jaw ſtrait and pointing forward, the two middlemoſt the longeſt : four tuſks ; thoſe in the upper jaw ſhort ; in the lower, very long and truncated obliquely ; ſometimes theſe teeth weigh ſix pounds nine ounces apiece, and are twenty-ſeven inches long *. The head of an enormous ſize : its mouth vaſtly wide : the ears ſmall and pointed, lined within very thickly with ſhort fine hairs : the eyes and noſtrils ſmall, in proportion to the bulk of the animal : on the lips are ſome ſtrong hairs, ſcattered in tufts, or *faſciculi*, here and there : the hair on the body is very thin, of a

* *Sparman* Stock: Wettſk. Handl. 1778. 329. tab.

3

whitiſh

whitish color, and scarce discernible at first sight: there is no mane on the neck, as some writers feign; only the hairs on that part are rather thicker: the skin is thicker even than that of a *Rhinoceros*, and of a dusky color: the tail is about a foot long, taper, depressed, and naked: the hoofs are divided into four parts; but, notwithstanding it is an amphibious animal, are unconnected by membranes: the legs short and thick.

In bulk, it is second only to the Elephant: the length of a male has been found to be seventeen feet; the circumference of its body fifteen; its height near seven; the legs near three; the head above three and a half; its girth near nine. Twelve oxen have been found necessary to draw one ashore, which had been shot in a river above the *Cape*. *Hasselquist* says, its hide is a load for a camel.

Inhabits the rivers of *Africa*, from the *Niger* to *Berg* river, many miles north of the Cape of *Good Hope*. These animals formerly abounded in the rivers nearer the *Cape*, but are now extirpated. To preserve the few which are left in *Berg* river, the governor has absolutely prohibited the shooting them, without particular permission.

It is not found in any of the *African* rivers which run into the *Mediterranean*, except the *Nile*, and even there only in the upper *Ægypt* *, and in the fens and lakes of *Æthiopia*, which that river passes through. Is a mild and gentle animal, unless it be

---

* Dr. *Shaw* says, that the present race of *Ægyptians* are not even acquainted with this animal; none ever appearing below the cataracts of the *Nile*. It was not so formerly; for *Radzivil* relates, that he saw and shot at four near *Damietta*. *Hasselquist* confirms the account of our countryman.

provoked:

provoked: inhabits equally the land and the water: swims very swiftly: during night leaves the rivers to graze: goes in troops sometimes six miles from the banks*, either in search of food or another river, and does great damage to the sugar-canes, and plantations of rice and millet: it also feeds on the roots of trees, which it loosens with its great teeth; but never eats fish. It is a clumsy animal on the land, walks slowly; but when pursued, takes to the water, plunges in, and sinks to the bottom, and is seen walking there at full ease: it cannot continue there long, it often rises towards the surface; but in the day time is so fearful of being discovered, that when it takes in fresh air, the place is hardly perceptible, for it does not venture even to put its nose out of the water. In rivers unfrequented by mankind, it is less cautious, and puts its whole head out of the water.

In shallow rivers it makes deep holes in the bottom, in order to conceal its great bulk. When it quits the water, it usually puts out half its body at once, and smells and looks around: but sometimes rushes out with great impetuosity, and tramples down every thing in its way.

Its voice is between the roaring of a bull and the braying of an elephant; and is at first interrupted with frequent short pauses. It may be heard at a great distance. **VOICE.**

If wounded, will rise and attack boats or canoes with great fury, and often sink them, by biting large pieces out of the sides: and frequently people are drowned by them; for they are as bold in the water, as they are timid on land. It is reported that they will at once bite a man in two. Are most numerous high

---

* *Journal historique*, 18.

up the rivers; frequently found near their mouths. It is now well known that they will at times enter the fea, not for the fake of feeding, but to fport for a time in greater expanfe. They will not even drink the falt water; but come on fhore in the night to quench their thirft in a neighboring well *.

They fleep in the reedy iflands in the middle of the ftream; and on which they bring forth their young. They perform the act of generation like our common cattle; and for that purpofe felect a fhallow part of the river.

They are capable of being tamed. *Belon* fays, he has feen one fo gentle, as to be let loofe out of a ftable, and fed by its keeper, without attempting to injure any one. They are generally taken in pit-falls; and the poor people eat the flefh, which is reckoned wholefome, and the fat is efteemed to be the beft lard. In fome parts, the natives place boards, full of fharp irons, in the corn-grounds; which thefe beafts ftrike into their feet, and fo become an eafy prey. Sometimes they are ftruck in the water with harpoons faftened to cords; and ten or twelve canoes are employed in the chafe †. The teeth are moft remarkably hard, even harder than ivory, and much lefs liable to grow yellow. It is certain that the dentifts prefer them for the making of falfe teeth. The fkin, when dried, is ufed to make bucklers, and is of an impenetrable hard-nefs.

A herd of females has but a fingle male: they bring one

* *Sparman*, ii. 285.

† *Purchas's Pilgr.* ii. 1544. *Haffelquift* gives an account of another method of taking them. The natives lay a great heap of peas in the places the *Hippopotame* frequents; it eats greedily, then growing thirfty, drinks immoderately; the peas in its belly fwell, the animal burfts, and is found dead. p. 188. *Engl. Ed.*

young

Female Hippopotame ___ 160

young at a time, and that on the land, but fuckle it in the water. Among other errors related of them, is that of their enmity with the *Crocodile*, an eye-witnefs declaring he had feen them fwimming together without any difagreement*.

Among the antient paintings in the *Rofpigliofi* palace, are fome moft ludicrous reprefentations of the chace of both thefe animals, by pygmies with long beards; and the fcenery fuitable. The painter, in the circumftance of the pygmies, dealt in the fiction of the times; in the former, fhewed his knowledge of the Hippopotame and Crocodile being joint tenants of the fame waters; and added the diminutive chaffeurs with much propriety, as they were faid by fome to have their refidence in the country of thofe tremendous animals.

It was known to the *Romans: Scaurus* treated the people with the fight of five *Crocodiles* and one *Hippopotame* †, during his ædilefhip; and exhibited them in a temporary lake. *Auguftus* produced one at his triumph over *Cleopatra* ‡. An antient writer afferts, that ‖ thefe animals were found in the *Indus*; which is not confirmed by any modern traveller.

This animal is the *Behemoth* of *Job*; who admirably defcribes its manners, food, and haunts.

    I. Behold now BEHEMOTH, which I made *near* thee; he eateth grafs as an ox.

    II. Lo! now, his ftrength is in his loins, and his force is in the navel of his belly.

    III. His bones are as ftrong pieces of brafs, his bones are like bars of iron.

    IV. He lieth under the fhady trees, in the covert of the reed and fens.

---

* *Purchas's Pilgr.* ii. 1544, 1568.   † *Plinii lib.* viii. c. 26.
    ‡ *Dion Caffius, lib.* li.   ‖ Vide *Gefner Pifc.* 419.

V. Behold! he drinketh up a river, and hasteth not. He trusteth he can draw up *Jordan* into his mouth.

The first, as the learned *Bochart* * observes, implies the locality of its situation, being an inhabitant of the *Nile*, in the neighborhood of *Uz*, the land of *Job*.

The second describes its great strength : and the third, the peculiar hardness of its bones.

The fourth, its residence, amidst the vast reeds of the river of *Egypt*, and other *African* rivers overshadowed with thick forests †.

The fifth, the characteristic wideness of its mouth : which is hyperbolically described as large enough to exhaust such a stream as *Jordan*.

* *Hierozoicon*, ii. 754.
† See *Masson's* travels, Ph. Transf. lxi. 292.

Tapir ___ 163. . N.83.

Fore hoofs divided into three parts; and a fort of falfe hoof behind.

Hind hoofs into three.

Tapiirete *Brafilienfibus, Lufitanis* Anta. *Marcgrave Brafil.* 229. *Pifo Brafil.* 101. *Nieuhoff's voy.* 23. *Raii fyn. quad.* 126. *Klein quad.* 36. Elephant hog. *Wafer's voy. in Dampier,* iii. 400. Mountain cow. *Dampier,* ii. 102. Sus aquaticus multifulcus. *Barrere France Æquin.* 160.

Anta ou grand Bete. *Gumilla Orenoque,* ii. 15. *Condamine voy.* 82. Species of Hippopotamus, or River Horfe. *Bancroft Guiana,* 127. Le Tapir ou Manipouris. *Briffon quad.* 81. *De Buffon,* xi. 444. *tab.* xliii. Hippopotamus terreftris. H. pedibus pofticis trifulcis. *Lin. fyft. Ed.* x. 74.

T. with the nofe extended far beyond the lower jaw; flender, and forming in the male a fort of probofcis, capable of being contracted or extended at pleafure; the fides fulcated; the extremities of both jaws ending in a point; ten cutting teeth in each; between them and the grinders, a vacant fpace: in each jaw ten grinders: ears erect, and oval, bordered with white: eyes fmall: body formed like that of a hog: the back arched: legs fhort: hoofs fmall, black, and hollow: tail very fmall: grows to the fize of a heifer half a year old: the hair is fhort: along the neck is a briftly mane, an inch and a half high: when young, is fpotted with white; when old, of a dufky color.

The nofe of the female is deftitute of the probofcis, and the jaws are of equal lengths.

Inhabits the woods and rivers of the *eaftern* fide of *South America,* from the ifthmus of *Darien* to the river of *Amazons:* fleeps,

Y 2

during

during day, in the darkeft and thickeft forefts adjacent to the banks: goes out in the night-time in fearch of food: lives on grafs, fugar-canes, and on fruits: if difturbed, takes to the water; fwims very well, or finks below, and, like the *Hippopotame*, walks on the bottom as on dry ground. The *Indians* fhoot it with poifoned arrows: they cut the fkin into bucklers, and eat the flefh, which is faid to be very good. Is a falacious, flow-footed, and fluggifh animal: makes a fort of hiffing noife.

MANNERS.

Thefe animals are of a very mild nature, and capable of being made very tame. In *Guiana* they are fometimes kept, and fed with other domeftic beafts in the farm-yards. They feed themfelves with their nofe, making ufe of it as the Rhinoceros does its upper lip. They know their mafter, who brings them their food: will take any thing that is offered, and rummage people's pockets with their nofe for meat. Their common attitude is fitting on the rump, like a dog*. Notwithftanding their mild nature, *Gumilla* fays, that, if attacked, they will make a vigorous refiftance; and fcarcely fails to tear off the fkin from the dogs which they can lay hold of.

Dampier and *Bancroft* give very faulty defcriptions of this beaft, imagining it to be the fame with the *Hippopotame*.

* *Allamand's* edit. of *De Buffon*, nouvelle ed. xv. 67. with two excellent figures.

No

Male Elephant___165. N.°84.

No cutting teeth; two vaſt tuſks; a long proboſcis.
Feet round, terminated by five ſmall hoofs.

Ελεφας. *Ariſt. Hiſt. An. lib.* i. *c.* 11. ix. *c.* 1.
Elephas. *Plinii lib.* viii. *c.* 1. *Geſner quad.* 376. *Raii ſyn. quad.* 131. *Klein quad.* 36. *Ludolph. Æthiop.* 54. *Boul-laye le Gouz,* 250. *Dellon's voy.* 71. *Leo Afr.* 336. *Kolben's Cape,* ii. 98. *Boſman's hiſt. Guinea,* 230. *Linſchot-ten iter,* 55. *Du Halde's China,* ii. 224. *Adanſon's voy.* 138. *Moore's* *trav.* 31. *Borri's account Cochin China,* 795. *Barbot's Guinea,* 141, 206, 207, 208. *Seb. Muſ.* i. 175. *tab.* iii. *Edw.* 221. *Schreber,* ii. 60. *tab.* lxxviii.
L'Elephant. *Briſſon quad.* 28. *De Buf-fon,* xi. 1. *tab.* i.
Elephas maximus. *Lin. ſyſt.* 48. *Fau-nul. Sinens. Br. Muſ. Aſhm. Muſ.* Lev. Mus.

E. with a long cartilaginous trunk, formed of multitudes of rings, pliant in all directions, terminated with a ſmall moveable hook: the noſtrils at the end of the trunk; its uſe that of a hand, to convey any thing into the mouth: no cutting teeth: four large flat grinders in each jaw; in the upper two vaſt tuſks, pointing forwards, and bending a little upwards; the largeſt * are ſeven feet long, and weigh 152 lb. each: the eyes ſmall: ears long, broad, and pendulous: back much arched: legs thick, and very clumſy and ſhapeleſs: feet undivided; but the margins terminated by five round hoofs: tail like that of a hog, terminated with a few long hairs, thick as packthread: color of the ſkin duſky, with a few ſcattered hairs on it.

The females have two teats, very ſmall in proportion to the bulk of the animal, and placed a little behind the fore legs.

* To be underſtood of thoſe imported into *England.*

The

SIZE.

The largeſt of land animals: there are certain accounts of their attaining the height of twelve feet; others are ſaid to have been three feet higher: but I ſuſpect that the laſt is exaggerated, and the firſt very rare. The height of nine feet and a half being reckoned a very tall beaſt.

PLACE.

Inhabits *India*, and ſome of its greater iſlands *, *Cochin-China*, and ſome of the provinces of *China*: abounds in the ſouthern parts of *Africa*, from the river *Senegal* to about two degrees north of the *Cape*†, and from thence as high as *Æthiopia* on the other ſide: found in greateſt numbers in the interior parts, where there are vaſt foreſts, near the ſides of rivers: are fond of marſhy places, and love to wallow in the mire like a hog: ſwim very well: feed on the leaves and branches of trees: do great damage to the fields of corn, and to plantations of *Coco Palms*, tearing up the trees by the roots to get at their tops.

MANNERS.

Often ſleep ſtanding; are not incapable of lying down, as is vulgarly believed; are very mild and harmleſs, except wounded, or during the rutting-time, when they are ſeized with a temporary madneſs: are ſaid to go nine months with young: this is gueſſed by the caſual eſcape of the tame females, when in rut, into the woods; where they couple with the wild: are ſoon diſcovered and brought back; and obſerved to bring forth in about

---

* *Soolo*, an iſland to the ſouth weſt of *Mindanao*, was deſtitute of elephants till a few were ſent as a preſent from *Siam*. Some eſcaped to the woods, and their offspring are now wild there. None are found in *Mindanao*, *Celebes*, or the other iſlands to the eaſt of *Scolo*. Capt. *Forreſt*.

† From the names of many places, it is probable that elephants were formerly found nearer to that great promontory; but at preſent none are ſeen further ſouth than the country of the *Amacquas*.

nine

nine months from the time. According to the *Ayeen Akbery**, i. 148, they are faid to go eighteen months. In a wild ftate the young elephants do not attach themfelves to their dams, but fuck indifferently the milch females of the whole herd. They bring only one at a time; very rarely two. The young are about three feet high when they are firft born; aad continue growing till they are fixteen or twenty years old. They are faid to live a hundred and twenty or thirty years †.

Drink by means of their trunk, fucking water up it, and then conveying it into the mouth; are very careful of the trunk, confcious that their exiftence depends on it; is to them as a hand; is their organ of feeling and of fmell, both which fenfes it has in the moft exquifite degree: its ftrength matchlefs; the tame elephants carry fmall pieces of artillery, fmall towers, with numbers of people in them, and alfo vaft loads: is not at prefent domefticated in *Africa*, only in the more civilized continent of *Afia*; they are much more numerous in *Africa*, in fome parts fwarm, fo that the negroes are obliged to make their habitations under ground for fear of them. Are killed and eaten by the natives; the trunk faid to be a delicious morfel: caught in pit-falls, covered with branches of trees; fometimes chaced and killed with lances: are inftantly killed by a flight wound in the head, behind the ears. All the teeth are brought from *Africa*; frequently picked up in the woods; uncertain whether fhed teeth, or from dead animals: the *African* teeth ‡, which come from *Mofambique*,

---

* Or inftitutes of the emperor *Akber*.

† *Tavernier*, ii. 96.

‡ *Dellon's voy.* 74. I have feen, in very large teeth, fmall brafs bullets lodged almoft in the centre: the orifice made by the ball was entirely filled up with the ivory matter, and the bullet formed a nucleus.

are.

are ten feet long; thofe of *Malabar* only three or four; the largeft in *Afia* are thofe of *Cochin-China*, which even exceed the elephants of *Mofambique* *. The fkin is thick, and, when dreffed, proof againft a mufket-ball : the flefh, the gall, the fkin, the bones, according to *Shi Chin*, are ufed in medicine †.

The wild elephants of *Ceylon* live in troops or families diftinct and feparate from all others, and feem to avoid the ftrange herds with particular care. When a family removes from place to place, the largeft-tufked males put themfelves at the head; and if they meet with a large river, are the firft to pafs it. On arriving on the oppofite bank, they try whether the landing-place is fafe : in cafe it is, they give a figna of a note from the trunk, as if it were the found of a trumpet, on which the remaining part of the old elephants fwim over; the little elephants follow, holding one another by locking their trunks together; and the reft of the old ones bring up the rear.

In the woods are often feen a folitary male elephant, wandering like an outlaw banifhed from the herd, and all the race. Thefe are as if in a ftate of defperation, and very dangerous. A fingle man will put to flight whole herds of focial elephants. This, alone, fears not his prefence, but will ftand firm, putting his power to defiance ‡.

MANNER OF
PASSING RIVERS.

* *Borri,* 795.

† *Du Halde China,* ii. 224.

‡ The feveral curious particulars inferted in this edition, refpecting the elephant, are taken from a memoir on the fubject, tranfmitted by Mr. *Marcellus Bles,* fecretary during twelve years to the *Dutch* government in *Ceylon,* and communicated to me by Governor LOTEN.

8

In

Female Elephant ———— *168.*

In *Ceylon* they are a great article of commerce, and are fold to the merchants of the *Indian* continent, who refort there to buy them for the ufe of the great men. This makes the taking of them a matter of importance. The *Ceylonefe* fometimes furround the woods where the elephants inhabit, with numerous bands, and drive before them, with all kinds of noifes, firing of guns, and with lighted torches, the beafts that happen to be there, till they are entrapped in a park inclofed with pallifades, conftructed in the foreft, in form of a wheel. At other times, the younger and moft active *Ceylonefe* follow them in the woods, and, putting them to flight, purfue till they have an opportunity of flinging a fort of fpringe, made of cord, round the hind legs of a beaft, which they follow, holding it in their hands till they can wind it round a tree : then they bring two tame elephants, which they place on each fide of the wild one, and fo conduct him home ; but fhould he prove reftive, they direct the tame ones to beat him with their trunks, which foon quiets even the moft ferocious.

A third way of taking the wild kind, is by means of tame female elephants, trained for the purpofe. Thefe the *Indians* carry into the woods, where the artful female foon enveigles a male out of the favage herd. As foon as fhe has made a conqueft, and feparated the male from his family, the *Indians* with a great noife terrify the reft, and put them to flight, and others make themfelves mafters of the beaft thus detached from its friends.

The report of the great fwiftnefs of the elephants is erroneous : an active and nimble *Indian* can eafily outrun them *.

* *M. Bles.* In *Borneo*, elephants are only found near a great inland lake, which feparates *Banjarmaling* from the empire of *Borneo*, and in no other part of the ifland. Their tufks are a great article of commerce.

By the obfervations of Mr. *Bles*, it is very long before the tufks arrive at a great fize: neither is it every male that has them of the magnitude we often fee; not one in ten has them, notwithftanding they may equal, in vigor and bulk of body, thofe which have: on the contrary, their tufks are fhort, flender, and blunt, and never above a foot long: nor is it poffible to know whether the tufks will be larger or not, till the beaft arrives at the age of twelve or fourteen.

SAGACITY.

Are, notwithftanding the great dullnefs of their eye and ftupidity of their appearance, the moft docile and moft intelligent of animals: tractable and moft obedient to their mafter's will: are fenfible of benefits, refentful of injuries: directed by a flight rod of iron hooked at one end: are in many parts of *India* the executioners of juftice; will, with their trunks, break every limb of the criminal, or trample him to death, or transfix him with their tufks, according as they are directed: are fo modeft as never to permit any one to fee them copulate: have a quick fenfe of glory. In *India*, they were once employed in the launching of fhips: one was directed to force a very large veffel into the water: the work proved fuperior to his ftrength: his mafter, with a farcaftic tone, bid the keeper take away this lazy beaft, and bring another: the poor animal inftantly repeated his efforts, fractured his fcull, and died on the fpot*. In *Delli*, an elephant paffing along the ftreets, put his trunk into a taylor's fhop, where feveral people were at work; one of them pricked the end with his needle: the beaft paffed on, but in the next dirty puddle filled its trunk with wa-

---

* *Ludolph. Com. in hift. Æthiop.* 147.

ter,

ter, returned to the fhop, and fpurting every drop among the people who had offended him, fpoilt their work *.

An elephant in *Adfmeer*, which often paffed through the *Bazar* or market, as he went by a certain herb-woman, always received from her a mouthful of greens: at length he was feized with one of his periodical fits of rage, broke his fetters, and, running through the market, put the crowd to flight; among others, this woman, who, in hafte, forgot a little child fhe had brought with her. The animal, recollecting the fpot where his benefactrefs was wont to fit, took up the infant gently in his trunk, and placed it in fafety on a ftall before a neighboring houfe †.

Another, in his madnefs, killed his *Cornac* or governor: the wife, feeing the misfortune, took her two children and flung them before the elephant, faying, *Now you have deftroyed their father, you may as well put an end to their lives and mine.* It inftantly ftopped, relented, took the greateft of the children, placed him on its neck, adopted him for its *Cornac*, and never afterwards would permit any body elfe to mount it ‡.

The *Indians* have from very early times employed the elephant in their wars: *Porus* oppofed the paffage of *Alexander*, over the *Hydafpes* ‖, with eighty-five of thefe animals: *M. de Buffon* very juftly imagines, that fome of the elephants which were taken by that monarch, and afterwards tranfported into *Greece*, were employed by *Pyrrhus* againft the *Romans*. From the time of *Solomon*, ivory has been ufed in ornamental works; it was one of the imports of his navy of *Tarfhifh*, whofe lading was gold and filver, ivory, apes, and peacocks §.

* *Hamilton's* account of *Eaft Indies*, ii. 109.   † *Terry's* Voyage, 148.
‡ *De Buffon*, xi. 77.   ‖ *Quint. Curtius*, lib. viii. c. 42.   § 2 *Chron.* ix. 21.

Z 2

An

An elephant was prefented, in 1254, to *Henry* III. by *Louis* IX. of *France* [*], which was kept with great care in the Tower. A writ was iffued to the Sheriffs of *London*, directing them to make fufficient provifion for *our* Elephant, *Elephans nofter*, and its keeper[†]; and another, which orders them to " build, out of " the city revenues, in our Tower of *London*, one houfe of forty " feet long and twenty deep, for our *Elephant* [‡]".

FOSSIL IN ENG-
LAND.

The teeth of this animal are often found in a foffil ftate; fome years ago two great grinding teeth, and part of the tufks of an elephant, were given me by fome miners, who difcovered them at the depth of forty-two yards, in a lead-mine in *Flintfhire*; one of the ftrata above them was lime-ftone, about eight yards thick; the teeth were found in a bed of gravel in the fame mine; the grinders were almoft as perfect as if juft taken from the animal; the tufk much decayed, foft, and exfoliating. A ftag's horn was found with them.

SIBERIA.

The grinders and tufks of the *Mammouth*, fo often found foffil in *Siberia*, muft be referred to this animal, as is evident from the account and figures of thofe in the *Ph. Tr. abridg.* ix. 87. by Mr. *Breynius* [||]. The *molares* differ not in the left from thofe recent; but the tufk has a curvature far greater than thofe of any elephant I have feen; whether this was accidental or preternatural, cannot be determined from a fingle fpecimen; *Strahlenberg* fays,

---

* *Matthew Paris*, 903.   † *Madox*, Antiq. Exch. i. 377.   ‡ *Maitland's London*, i. 171.

|| Who has given very accurate figures of the entire head, the *molares*, the tufk, and the thigh bone.

they

they are fomewhat more crooked * than elephants teeth commonly are; and others relate that a pair weighed 400 lb, which
exceeds the weight of the largeft recent tufks: there are alfo
found with them foffil grinders of 24 lb. weight; but fince, in all
other refpects, thofe grinders refemble thofe of the living elephants; and one being found lodged in the fkeleton of the fame
head with the tufks, we cannot deny our affent to the opinion of
thofe who think them to have been once the parts of the animal
we have juft defcribed.

Entire fkeletons, or parts of them, teeth, and tufks, are found
in prodigious quantities all over northern *Afia*, there not being
the bank of any great river in which they are not met. with,
wafhed out of the clay or rather muddy ftrata, in which they
are lodged. All the country towards the *Arctic* circle is a vaft
moffy flat, formed of a bed of mud or fand, feeming the effect
of the fea, and which gives great reafon to think, that immenfe
tract was in fome very diftant age won from it. With them are
mixed an infinitely greater number of marine bodies, than are
found in the higher parts of that portion of *Afia*. I give the fact:
let others, more favored, explain the caufe how thefe animals
were tranfported from their torrid feats to the *Arctic* regions, for
(as I have before mentioned, that the Rhinoceros, and the Antelope, have been found at this diftance from their native
country, a flood muft have brought them here, and a fudden
retreat of the water left them) I fhould have recourfe to the
only one we have authority for: and think that phænomenon
fufficient: I mention this, becaufe modern philofophers look out

* *Hift. Ruffia,* 402. Alfo *Bell's Travels,* ii. 165. *Le Brun's Travels,* i. 63.

for

for a later caufe: I reſt convinced; therefore avoid contradict-
ing what never can be proved.

The tuſks are made uſe of as ivory, formed into combs, and
uſed to inlay cabinets: and are a great article of commerce, eſpecially with the *Chineſe.* The *Tartars* have many wild notions
about the *Mammouth*, ſuch as its being a ſubterraneous animal,
&c. &c. *Linnæus** ſays, it is the ſkeleton of the *Walrus* flung
on ſhore.

An animal only known in a foſſil ſtate, and that but partially;
from the teeth, ſome of the jaw-bones, the thigh-bones, and *ver-
tebræ*, found with many others five or ſix feet beneath the ſurface, on the banks of the *Ohio*, not remote from the river *Miame*,
ſeven hundred miles from the ſea-coaſt.

Some of the tuſks near ſeven feet long, one foot nine inches
in circumference at the baſe, and one foot near the point; the
cavity at the root or baſe nineteen inches deep: the tuſks of the
true elephant have ſometimes a very ſlight lateral bend, theſe have
a larger twiſt or ſpiral curve towards the ſmaller end; but the
great and ſpecific difference conſiſts in the ſhape of the *molares* or
grinders, which are made like thoſe of a carnivorous animal, not
flat, and ribbed tranſverſely on their ſurface, like thoſe of the recent elephant, but furniſhed with a double row of high and conic
proceſſes, as if intended to maſticate, not to grind their food.

A third difference is in the thigh-bone: which is of a great
diſproportionable thickneſs to that of the elephant, and has alſo
ſome anatomical variations.

The tuſks have been cut and poliſhed by the workers in ivory,

85. AMERICAN.

USES.

* *Syſt. Nat.* 49.

who

who affirmed, that in texture and appearance they differed not from the true ivory: the *molares* were indurated to a great degree. Specimens of thefe teeth and bones are depofited in the *Britiſh Muſeum*, that of the *Royal Society*, and in the cabinet of that liberal man the late Doctor *Hunter* *. I ſhould have been leſs accurate in this defcription, had not that gentleman favored me with his obſervations on fome particulars, which otherwiſe might have eſcaped my notice.

Theſe foſſil bones are alſo found in *Peru*, and in the *Brazils*: as yet the living animal has evaded our ſearch; it is more than probable that it yet exiſts in fome of thoſe remote parts of the vaſt new continent, unpenetrated yet by *Europeans*. Providence maintains and continues every created ſpecies; and we have as much aſſurance, that no race of animals will any more ceaſe while the earth remaineth, *than ſeed-time and harveſt, cold and heat, ſummer and winter, day or night*.

Theſe reliques are not peculiar to *America*, for fome have of late years been difcovered in *Siberia*, and perhaps in *Ruſſia* †. It is remarked, that they are not only met with more rarely than thofe of the true elephants, but even at greater depths: in fuch ſtrata, which are fuppofed to have been the ruins of the old world, after the event of the deluge.

To this may properly be added a very obfcure animal, mentioned by *Nieuhoff* ‡, and called by the *Chineſe* of *Java Suko-*

SIBERIA.

---

* Who has obliged the world with an ingenious eſſay on the ſubject, *vide Ph. Tr.* vol. lviii. 34. The late worthy *Peter Collinſon*, in the preceding volume, gave us other notices of thefe bones.

† *Pallas* in Act. Acad. *Petrop.* ii. 219;

‡ *Nieuhoff's voy. in Churchill's coll.* ii. 360.

tyro. It is of the size of a large ox : has a snout like a hog : two long rough ears ; and a thick bushy tail : the eyes placed upright in the head, quite different from other beasts : on the side of the head, next to the eyes, stand two long horns, or rather teeth, not quite so thick as those of an elephant. It feeds on herbage, and is but seldom taken.

D I V.

DIV. II.

DIGITATED QUADRUPEDS.

D I V. II.   Digitated Quadrupeds.

S E C T. I. Anthropomorphos *.

XV. APE.

Four cutting teeth in each jaw, and two canine.

Each of the feet formed like hands, generally with flat nails, and, except in one inftance, have four fingers and thumb.

Eye-brows above and below.

A Moft numerous race; almoft confined to the torrid zone: fills the woods of *Africa*, from *Senegal* to the *Cape*, and from thence to *Æthiopia:* a fingle fpecies is found beyond that line, in the province of *Barbary:* found in all parts of *India*, and its iflands; in *Cochin-China*, in the S. of *China*, and in *Japan*; and one kind is met with in *Arabia:* they fwarm in the forefts of *South America*, from the ifthmus of *Darien*, as far as *Paraguay*.

Are lively, agile, full of frolic, chatter, and grimace: from the ftructure of their members, have many actions in common

---

* *Animals approaching the human form:* A term to be taken in a limited fenfe; to be applied to all of this fection, as far as relates to their feet, which ferve the ufes of hands in eating, climbing, or carrying any thing: to the flatnefs of the nails, in many fpecies; and to fome refemblance of their actions, refulting from the ftructure of their parts only, not from any fuperior fagacity to that of moft others of the brute creation.

with

with the human kind: moſt of them are fierce and untameable;
ſome are of a milder nature, and will ſhew a degree of attachment; but in general are endowed with miſchievous intellects:
are filthy, obſcene, laſcivious, thieving: feed on fruits, leaves,
and inſects: inhabit woods, and live in trees; in general are gregarious, going in vaſt companies: the different ſpecies never
mix with each other, always keep apart and in different quarters:
leap with vaſt activity from tree to tree, even when loaded with
their young, which cling to them. Are the prey of leopards,
and others of the feline race; of ſerpents, which purſue them to
the ſummit of the trees, and ſwallow them entire. Are not carnivorous; but for miſchief-ſake will rob the neſts of birds of the
eggs and young. In the countries where apes moſt abound, the
ſagacity of the feathered tribe is more marvellouſly ſhewn in
their contrivances to fix their neſt beyond the reach of theſe in-
vaders *.

Apes and parrots (the apes of birds) are more numerous in
their ſpecies than any other animals; their numbers and their
different appearances made it neceſſary to methodize and ſubdi-
vide the genus; accordingly Mr. *Ray* firſt diſtributed them into
three claſſes:

Simiæ, APES, ſuch as wanted tails.

Cercopitheci, MONKIES, ſuch as had tails.

And from the laſt he formed another diviſion, viz.

Papiones, BABOONS, thoſe with ſhort tails; to diſtinguiſh
them from the common monkies, which have very long ones. I
comprehend in this diviſion of baboons, ſuch whoſe tails do not

* *Indian Zoology, p. 7. tab.* viii.

A a 2   exceed

exceed half the length of their bodies, and commonly carried in an arched direction. Heads large; bodies short.

From *Ray, Linnæus* formed his method; *M. de Buffon* followed the same; but makes a very judicious subdivision of the long-tailed apes, or the true monkies, into such which had prehensile tails*, and such which had not. I shall endeavor in this genus no other reform in the system of our countryman, than what that gentleman has made; in respect to the *trivial* names of the species, I have in general invented such as I supposed congruous, or in a few instances retained those of *M. de Buffon*.

*

Without tails; the true APES.

86. GREAT.

Satyrus. *Gesner quad.* 863.
Pongo. *Purchas's Pilgr.* ii. 982. v. 623.
Homo sylvestris, orang outang. *Bontius Java*, 84. *Beckman's Borneo*, 37.
Baris. *Nieremberg*, 179.
Barrys. *Barbot's Guinea*, 101.
Quojas morrou. *idem.* 115.
Chimpanzee. *Scotin's print*, 1738.
Man of the wood. *Edw.* 213.
Le Jocko. *de Buffon*, xiv. 44. *tab.* i.
Le Pongo. *ibid.*

L'Homme de bois. Simia unguibus omnibus planis et rotundatis cæsarie faciem cingente. *Brisson quad.* 134.
Homo Troglodytes. Homo nocturnus. *Lin. syst.* 33. *Amœn. Acad.* vi. 63. 69. 72.
Simia satyrus. S. ecaudata ferruginea, lacertorum pilis reversis, natibus tectis. *Lin. syst.* 34. *Br. Mus.*
L'orang outang. *Schreter*, 64. *tab.* i. ii.

A with a flat face, and a deformed resemblance of the human: ears exactly like those of a man: hair on the head longer than on the body: body and limbs covered with reddish and

---

* Animals with this kind of tail can lay hold of any thing with it, for it serves all the uses of a hand; they can twist it round the branch of a tree, and suspend themselves by it, or keep them secure in their seat, while their feet are otherwise employed. This faculty is common to some *Monkies*, to *Macaucos*, and one species of *Porcupine*. *Vide* plates of yellow *Macauco*, and *Brasilian Porcupine*.

shaggy

Ourang Outang or Great Ape — Nº 86

ſhaggy hair; longeſt on the back, thinneſt on the fore-parts: face and paws ſwarthy: buttocks covered with hair.

This ſeems the leſſer kind, and is that engraven by Mr. *Edwards*, tab. 213, and by Mr. *Schreber*, tab. 1.

The *Pongo* of *Purchas* is the greater, more robuſt, muſcular, of a deeper color, and very thinly furniſhed with hair. This is figured by *de Buffon*, xiv. tab. i. and by *Schreber*, tab. ii. The hiſtory of theſe is ſtill obſcure, nor are we aſſured whether they are diſtinct ſpecies or only varieties.

Inhabit the interior parts of *Africa* and the iſle of *Borneo*. Are ſolitary, and live in the moſt deſert places: grow to the height of ſix feet: have prodigious ſtrength; will overpower the ſtrongeſt man. The old ones are ſhot with arrows; only the young can be taken alive. Live entirely on fruits and nuts: will attack and kill the negroes who wander in the woods: will drive away the elephants, and beat them with their fiſts, or pieces of wood: will throw ſtones at people that offend them: ſleep in trees; make a ſort of ſhelter from the inclemency of the weather: are of a ſolitary nature, grave appearance, and melancholy diſpoſition, and even when young not inclined to frolic: are vaſtly ſwift and agile: go erect: ſometimes carry away the young negroes*.

When taken young are capable of being tamed: very docile; are taught to carry water, pound rice, turn a ſpit. The *Chimpanzee* ſhewn in *London*, 1738, was extremely mild, affectionate,

---

* Theſe accounts are chiefly taken from *Andrew Battel*, an *Engliſh* ſailor, who was taken priſoner 1589, and lived many years in the inner parts of *Congo*; his narrative is plain, and ſeems very authentic. It is preſerved in *Purchas's* collection.

good-

good-natured; like the fatyr of *Pliny, mitiſſima natura*; very
fond of the people it was uſed to: would eat like a human creature: lay down in bed like one, with its hand under its head:
fetch a chair to fit down on: drink tea, pour it into a faucer if
too hot: would cry like a child; be uneafy at the abſence of its
keeper. This was only two feet four inches high, and was a
young one: that deſcribed by Doctor *Tyſon* * two inches ſhorter.
There is great poſſibility that theſe animals may vary in ſize and in
color, ſome being covered with black, others with reddiſh hairs.

Not the *Satyrs* of the antients, which had tails †, and were a
ſpecies of monkey. *Linnæus's Homo nocturnus*, an animal of this
kind, unneceſſarily ſeparated from his *Simia Satyrus*. Some of
the authorities in the *Amœn. Acad.* very doubtful. Sir *John Man-*
deville, p. 361, certainly meant this large ſpecies, when he ſays
he came to *another yle where the Folk ben alle ſkynned roughe heer,*
as a rough beſt, ſaf only the face, and the pawme of the hond.

---

* *Orang outang, five homo ſylveſtris*; or the anatomy of a Pygmy. *Folio.*
London. 1699.

† *Ælian* gives them tails, *lib.* xvi. *c.* 21. *Pliny* ſays they have teeth like dogs,
lib. vii. *c.* 2. circumſtances common to many monkies. *Ptolemy, lib.* vii. *c.* 2.
ſpeaks of certain iſlands in the *Indian* ocean, inhabited by people with tails like
thoſe with which *Satyrs* are painted, whence called the iſles of *Satyrs*. *Kœping,*
a *Swede,* pretended to have diſcovered theſe *Homines Caudati*; that they would
have trafficked with him, offering him live parrots; that afterwards they killed
ſome of the crew that went on ſhore, and eat them, &c. &c. *Amœn. Acad.* vi. 71.

Πιθηκος.

Πιθηκος. *Ariſtot hiſt an. lib. c.* 8.
Simia. *Geſner quad.* 8_47. *Raii ſyn. quad.*
  149.
Ape, 2d. ſp. *Boſman's Guinea.* 242.
Le Singe. Simia unguibus omnibus pla-

nis et rotundatis. *Briſſon quad.* 133.
Le Pitheque. *de Buffon*, xix. 84.
Simia ſylvanus. S. ecaudatus, natibus
  calvis capite rotundato. *Lin. ſyſt.* 34.
Le ſinge commun. *Schreber*, 80. *tab.* iv.

87. PIGMY.

A. with a flattiſh face : ears like thoſe of a man : body of the ſize of a cat : color above of an olive brown, beneath yellowiſh : nails flat : buttocks naked : ſits upright.

Inhabits *Africa*. Not uncommon in our exhibitions of animals : very tractable, and good-natur'd : moſt probably the pygmy of the antients. Abounds in *Æthiopia**, one ſeat of that imaginary nation : was believed to dwell near the fountains of the *Nile* †; deſcended annually to make war on the cranes, i. e. to ſteal their eggs, which the birds may be ſuppoſed naturally to defend; whence the fiction of their combats. *Strabo* judiciouſly ‡ obſerves, that no perſon worthy of credit ever ventured to aſſert he had ſeen this nation : *Ariſtotle* ſpeaks of them only by hearſay, ωσπερ λεγεται : they were ſaid to be mounted on little horſes, on goats, on rams, and even on partridges. The *Indians* taking advantage of the credulity of people, embalmed this ſpecies of ape with ſpices, and ſold them to merchants as true pygmies § : ſuch, doubtleſs, were the diminutive inhabitants mentioned by Mr. *Groſe* ‖ to be found in the foreſts of the *Carnatic*.

* *Ludolph. Æthiop.* 57.
† *Ariſt. hiſt. an. lib.* viii. *c.* 13.
‡ *Geſner quad.* 852, from *Marco Polo.* They take off all the hair, except a
little by way of beard.
§ *Lib.* xvii.
‖ *Voy. E. Indies,* 365.

5

Feed

FOOD.   Feed on fruits; are very fond of infects, particularly of ants: affemble in troops *, and turn over every ftone in fearch of them. If attacked by wild beafts, take to flight; but if overtaken, will face their purfuers, and by flinging the fubtile fand of the defert in their eyes, often efcape †.

88. LONG-ARMED.   Le grand Gibbon. *de Buffon,* xiv. 92. *tab.* ii. *Schreber,* 78. *tab.* iii.

A. with a flat fwarthy face, furrounded with grey hairs: hair on the body black and rough: buttocks bare: nails on the hands flat; on the feet long: arms of a moft difproportioned length, reaching quite to the ground when the animal is erect, its natural pofture: of a hideous deformity. Grows to the height of four feet: fometimes walks upright; fometimes on all fours.

α LESSER. Refembling the former, but much lefs: its colors brown and grey. From *Malacca.* Le petit gibbon. *de Buffon,* xiv. *tab.* iii. *Schreber,* 80. *tab.* iii. *f.* 2. MUS. LEV.

β A fpecies in poffeffion of Lord *Clive,* about two years ago, much refembling the laft, but more elegant in its form, and the arms fhorter; but fo nearly allied in fhape, as not to be feparated: face, ears, crown of the head, feet, and hands, black: the reft of the body and arms covered with filvery hairs: about three feet high: good-natured, and full of frolic. That which

* *Ludolph. Æthiop.* 57.
† *Idem,* 58.

we

Long armed Ape. ___ N.88.

we have engraven is in the *Leverian* Museum; and is remarkable
for the great length and shagginess of its hair: seemingly needless
for a native of the torrid zone. It was a female, and not three feet
high.

These animals are mild, gentle, and modest; feed on
leaves, fruits, and barks of trees. Inhabit *Malacca*, the *Molucca*
islands, and *Sumatra*, where they are to be seen by hundreds on the
tops of trees *. These last seem our lesser variety, not exceeding
three feet in height. They walk erect, and never on all four.

The great black ape of *Mangsi*, a province of *China*, is probably of this kind †.

Ph. Transf. lix. 72. tab. iii.

A with a pointed face, long and slender limbs: arms, when the
animal is upright, do not reach lower than the knees: head
round, and full of hair: grows to the height of a man.

Inhabits the forests of *Mevat*, in the interior parts of *Bengal*.
They are gentle and modest, called by the natives *Golok*, or wild
men: distinct from the *orang outang*, by their slender form; from
the *Gibbons*, by having shorter arms.

Simia Lar. Gm. Lin. 27.   Miller's plates, tab. xxvii.

A with black face, crown of the head, fingers, and inside of the
feet and hands: round the face the hair long and whitish, on
the cheeks and chin forming a beard: hair on the body short and

* Phil. Transf. vol. lxviii. part. i. 170.   † Du Halde China, i. 118.

VOL. I.   B b   dusky:

dufky: limbs very long: face obtufe.   A fmall fpecies: feemingly diftinct from the others.

Inhabits, according to Mr. *Miller, China.*

91. BARBARY.

Κυνοκεφαλος. *Ariftot. hift. an. lib.* ii. *c.* 8.
Cynocephalus. *Plinii lib.* viii. *c.* 54.
Simius cynocephalus. *Pr. Alp. Ægypt.* i. 241. *tab.* xv. xvi.
Le Magot. *de Buffon,* xiv. 109. *tab.* vii. viii. *Shaw, Spec. Lin.* i.
Le Singe Cynocephale. *Briffon quad.*

135. *Schreber,* 84. *tab.* v.
Simia Inuus. S. ecaudata natibus calvis, capite oblongo. *Lin. fyft.* 35.
Yellow ape? *Du Halde China,* i. 120.
La Roque voy. Arabie, 210. Mus. LEV.

A with a long face, not unlike that of a dog: canine teeth, long and ftrong: ears like the human: nails flat: buttocks bare: color of the upper part of the body a dirty greenifh brown: belly of a dull pale yellow: grows to above the length of four feet.

PLACE.

Inhabits many parts of *India, Arabia,* and all parts of *Africa,* except *Ægypt,* where none of this genus are found. A few are found on the hill of *Gibraltar,* which breed there: probably from a pair that had efcaped from the town; for I never heard that they were found in any other part of *Spain.*

MANNERS.

Are very ill-natured, mifchievous, and fierce; agreeing with the character of the antient *Cynocephali:* are a very common kind in exhibitions: by force of difcipline, are made to play fome tricks; otherwife, are more dull and fullen than the reft of this genus: affemble in great troops in the open fields in *India**: will attack women going to market, and take their provifions

* *Dellon's voy.* 83.

from

1. *Hog-faced Baboon ___ N.º 92.*

2. *Brown Baboon ___ N.º 99.*

from them. The females carry the young in their arms, and will leap from tree to tree with them. Apes were worſhipped in *India*, and had magnificent temples erected to them. When the *Portugueſe* plundered one in *Ceylon*, they found in a little golden caſket * the tooth of an ape; a relique held by the natives in ſuch veneration, that they offered 700,000 ducats to redeem it, but in vain; for it was burnt by the Viceroy, to ſtop the progreſs of idolatry.

A. With ſhort tails.

92. HOG-FACED.

Ariſtotle barely mentions another ſpecies of ape, under the title of χοιροπιθηκος, *ſimia Porcaria*. In tab.   of this work is an engraving of this animal, taken from the drawing of one in the *Britiſh Muſeum*, with a noſe exactly reſembling that of a hog, which poſſibly may be *Ariſtotle's* animal; but there is no account attending the painting, to enable us to trace its hiſtory.

M. Gmelin, in his *Syſt. Lin.* refers to *Boddaert* Naturf. 22. p. 17. tab. i. ii. ſays it is half-tailed, has a naked face, olive brown body; buttocks cover'd, nails ſharp. In ſize of the length of three feet ſix inches.

Inhabits *Africa*.

* *Linſchottan's voy.* 53. In *Amadabat* are hoſpitals for apes, and other maimed animals. *Tavernier's voy. part.* ii. 48. The ſame writer ſays, that they breed in great numbers in *India*, in the copſes of *Bamboos*, which grow on each ſide the road, *p.* 94.

93. GREAT.   Papio. *Gefner quad.* 560, with a good   Le Choras. Simia mormon. *Alſtroemer,*
               figure.   *Schreber,* 92. tab. viii. MUS. LEV.
               Simia ſphynx. *Lin. ſyſt.* 35.

B. with hazel irides: ears ſmall and naked: face canine, and very thick: middle of the face and forehead naked, and of a bright vermilion color; tip of the noſe of the ſame: it ended truncated like that of a hog: ſides of the noſe broadly ribbed, and of a fine violet hue: the opening of the mouth very ſmall: cheeks, throat, and goat-like beard yellow: hair on the forehead is very long; turns back, is black, and forms a kind of pointed creſt. Head, arms, and legs covered with ſhort hair, yellow and black intermixed; the breaſt with long whitiſh yellow hairs; the ſhoulders with long brown hair.

Nails flat; feet and hands black: tail four inches long, and very hairy; buttocks bare, red, and filthy, but the ſpace about them is of a moſt elegant purple color, which reaches to the inſide of the upper part of the thighs.

This was deſcribed from a ſtuffed ſpecimen in Sir ASHTON LEVER's Muſeum. In *October,* 1779, a live animal of this ſpecies was ſhewn at *Cheſter,* which differed a little in color from the above, being in general much darker. Eyes much ſunk in the head, and ſmall. On the internal ſide of each ear was a white line, pointing upwards. The hair on the forehead turned up like a toupée. Feet black: in other reſpects reſembled the former.

In

Great Baboon — N.º 93.

Great Baboon — N.º 93.

P. Mazell Sculp.

In this I had opportunity of examining the teeth: the cutting-teeth were like thofe of the reft of the genus; but in the upper and lower jaw were two canine, or rather tuſhes, near three inches long, and exceedingly ſharp and pointed. This makes me fub-fcribe to Mr. *Schreber*'s opinion, that the TUFTED APE of my former edition was defignedly cropped and difguifed by its keeper, to render it a monfter *. I offer in my defence of hav-ing inferted it as a genuine fpecies, that it had been defcribed by Doctor *Bradley*, and adopted by the Royal Society, and placed in their inftructive Tranfactions.

This animal was five feet high, of a moft tremendous ſtrength in all its parts; was exceffively fierce, libidinous, and ſtrong. SIZE.

Mr. *Schreber* fays, that this fpecies lives on fucculent fruits, and on nuts: is very fond of eggs, and will put eight at once into its pouches; and, taking them out one by one, break them at the end, and fwallow the yolk and white. Rejects all fleſh-meat, unlefs it be dreffed: would drink quantities of wine or brandy. Was lefs agile than other baboons: very cleanly, for it would immediately fling its excrements out of its hut. MANNERS.

That which was ſhewn at *Chefter* was particularly fond of cheefe. Its voice was a kind of roar, not unlike that of a lion, but low and fomewhat inward. It went upon all fours, and never ftood on its hind legs, unlefs forced by the keeper; but would frequently fit on its rump in a crouching manner, and drop its arms before the belly. I have given a figure of that in the LEVERIAN MUSEUM, and another taken from the live animal, which ſhews its common and natural attitude. The laft will be

* I leave the figure as copied from the drawing in the *Britiſh Mufeum.*

a proof.

a proof of the excellence of *Gefner's* * figure of this fpecies, hitherto thought erroneous.

PLACE.   Inhabits the hotter parts of *Africa.*

94. RIBBED NOSE. Le Mandrill. *de Buffon,* xiv. 154. *tab.*   nis cæruleis ftriatis. *Lin. fyft.* 35.
            xvi. xvii.   Le Maimon. *Schreber,* 90. *tab.* vii.
            S. maimon. S. caudata fubbarbata ge-   *Shaw, Spec. Lin.* 2.

B. with a long naked nofe compreffed fideways, of a purple color, and ribbed obliquely on each fide: on the chin a fhort, picked, orange beard: tail very hairy, about two inches long, which it carries erect: buttocks naked: hair foft, dufky mottled with yellow: length from nofe to tail, about two feet.

PLACE.   Inhabits *Guinea.* Thofe I have feen fat erect on their rump, but walked on all fours: were good-natured, but not fportive.

Linnæus places this among the *fimiæ cauda elongata,* and applies to it fome of the fynonyms of the 72d fpecies: but his defcription agrees with this fo exactly, that there can be no doubt but that it is his *Simia maimon.*

This animal is well defcribed by M. *de Buffon,* Mr. *Ray,* *Linnæus,* and M. *Briffon;* and indeed every Naturalift, except M. *de Buffon,* has copied *Gefner:* but we think the firft ought to have applied the name of Baboon to this fpecies, inftead of that defcribed by him, p. 133; the one having the character of this fection, the other having a length of tail, that conftitutes that of the monkey.

The animal called, by *Barbot* and *Bofman* †, SMITTEN, is a

* P. 560. *Papio,* a barbarous name: from whence the *Englifh,* Baboon: *Italian,* Babbuino; and *French* Babouin.

† *Barbot's Guinea,* 212. *Bofman's Guinea,* 242.

4   large

Wood Baboon —— N.º 95.

large species to be referred to this genus: it is described with a great head, short tail, and of a mouse color; that it grows to the size of five feet, is very fierce, and will even attack a man.

The *Tretretretre of Madagascar* is another animal of this kind; described to be of the size of a calf of two years old; to have a round head, visage and ears of a man, feet of an ape, hair curled, very short tail; a solitary species: the natives are greatly afraid of it, and fly its haunts as it does theirs *.

B. with a long dog-like face, covered with a small glossy black skin: hands and feet naked, and black like the face: hair on all parts long, elegantly mottled with black and tawny; nails white.   95. WOOD.

About three feet high when erect: tail not three inches; and very hairy on the upper part.

Inhabits *Guinea*, where it is called by the *English*, the *Man of the Wood*. LEV. MUS.

B. with a black long face: ears hid in the fur: over the eyes are several long dusky hairs: hands covered above with hair: color a bright yellow, mottled with black. This greatly resembles the wood Baboon, except in size, and its hairy hands.   96. YELLOW.

These two are about two feet long: probably natives of *Africa*; but their place, age, and history obscure. LEV. MUS.

B. with a dusky face: pale brown beard: body and limbs of a cinereous brown: crown mottled with yellow. LEV. MUS.   97. CINEREOUS.

* *Flacourt, hist. Madag.* 154.

B. with

98. BROAD-
TOOTHED.

B. with a blueish face: two very flat broad fore teeth: a pale brown beard: long hairs over each eye: a tuft of hair beyond each ear; the hair black and cinereous, mixed with dull ruſt-color.

Length about three feet. A fuller hiſtory of theſe three is wiſhed. LEV. MUS.

99. BROWN.

Simia Platypygos. *Schreber*, 89. tab. v. B.

B. with pointed ears, face of a dirty white; noſe large and broad: hairs round the face ſhort and ſtrait: color of the upper part of the body brown; of the under, aſh-color.

Tail about four inches long; taper, and almoſt bare of hair. Beneath, is quite naked. The animal which I called the New Baboon, in the firſt edition, ſeems by the taperneſs of the tail, and general form, to be of this kind.

100. LITTLE.

Simia apedia. S. ſemicaudata, palmarum pollice approximato, unguibus oblongis pollicum rotundatis, natibus tectis. *Lin. ſyſt.* 35.

Simia cauda abrupta, unguibus compreſſis obtuſiuſculis, pollice palmarum digitis adhærente. *Amæn. Acad.* i. 558.

B. with a roundiſh head, mouth projecting, ears roundiſh, and naked; thumb not remote from the fingers: nails narrow, and compreſſed; thoſe of the thumbs rounded: color of

the

the hair yellowiſh, tipt with black: face brown, with a few ſcattered hairs: tail not an inch long: buttocks covered with hairs: ſize of a ſquirrel, according to *Linnæus.* But Mr. *Balk,* in the *Amœn. Acad.* ſays it is as large as a cat.

Inhabits *India:* is a lively ſpecies.

B. with the hairs on the crown very long, and diſhevelled; thoſe on the cheeks of the ſame form, and of a duſky color; breaſt whitiſh: reſt of the body and limbs covered with black long hair. Face and feet black and bare: tail ſlender, taper, about ſeven inches long: whole length of the animal two feet.

Inhabits *Africa.* Lev. Mus.

101. Crested.

Pig-tailed Monkey. *Edw.* 214.
Le Maimon. *de Buffon,* xiv. 176. *tab.* xix.
Simïa Nemeſtrina. S. Semicaudata ſub-

barbata griſea iridibus brunneis, natibus calvis. *Lin. ſyſt.* 35. *Br. Muſ.* Lev. Mus.

102. Pig-tail.

B. with a pointed face, not ſo long as that of the laſt: eyes hazel: above and beneath the mouth ſome few black hairs: face naked, of a ſwarthy redneſs: two ſharp canine teeth: ears like the human: crown of the head duſky: hair on the limbs and body brown, inclining to aſh-color, paleſt on the belly: fingers black: nails long and flat: thumbs on the hind feet very long, connected to the neareſt toe by a broad membrane: tail four inches long, ſlender, exactly like a pig's, and almoſt naked : the bare ſpaces on the rump red, and but ſmall : length, from head to tail, twenty-two inches.

Inhabits the iſle of *Sumatra* and *Japan* * : is very docile : in *Japan* is taught ſeveral tricks, and carried about the country by mountebanks. *Kæmpfer* was informed by one of theſe people, that the Baboon he had was 102 years old.

B A B O O N S.

** With longer tails.

| 103. Dog-faced. | Le Tartarin. *Belon, portraits, 102.*
Simia Ægyptiaca cauda elongata, clunibus tuberoſis nudis. *Haſſelquiſt, itin.* 189.
Simia Hamadryas. S. caudata cinerea, auribus comoſis, unguibus acutiuſculis, natibus calvis. *Lin. ſyſt. 36.* | Cercopithecus cynocephalus, parte anteriore corporis longis pilis obſita, naſo violaceo nudo. Le Magot ou le Tartarin. *Briſſon quad.* 152. *Edw. fig. ined.*
Le Babouin gris. *Schreber*, 100. tab. x.
Shaw, Spec. Lin. iii. |

B. with a long, thick, and ſtrong noſe, covered with a ſmooth red ſkin : eyes ſmall : ears pointed, and hid in the hair : head great, and flat : hair on the ſides of the head, and fore-part of the body, as far as the waiſt, very long and ſhaggy ; grey and olive brinded ; that on the top and hind part of the head very ſhort : the hair on the limbs and hind part of the body alſo ſhort : limbs ſtrong and thick : hands and feet duſky : the nails on the fore feet flat ; thoſe on the hind like a dog's : buttocks very

* *Kæmpfer's hiſt. Japan;* i. 126.

bare,

1 *Dog fac'd Baboon* — . No 103
2. *Purple fac'd . Monkey* — . No 107

bare, and covered with a skin of a bloody color : tail scarcely the length of the body, and carried generally erect.

Inhabit the hottest parts of *Africa* and *Asia*: keep in vast troops: are very fierce and dangerous: rob gardens: run up trees when passengers go by: shake the boughs at them with great fury, and chatter very loud : are excessively impudent, indecent, lascivious: most detestable animals in their manners, as well as appearance. Mr. *Edwards* communicated to me an account and a fine print * of one, which was shewn in *London* some years ago: it came from *Mokha*, in the province of *Yeman*, in *Arabia Felix*. They inhabit the woods by hundreds, which obliges the owners of the coffee-plantations to be continually on their guard against their depredations †. This animal was above five feet high; very fierce, and untameable; so strong, as easily to master its keeper, a strong young man : its inclinations to women appeared in the most violent manner. A footman, who brought a girl to see it, in order to teize the animal, kissed and hugged her: the beast, enraged at being so tantalized, caught hold of a quart pewter pot, which he threw with such force, and so sure an aim, that had not the man's hat and wig softened the blow, his scull must have been fractured; but he fortunately escaped with a common broken head.

Of the same kind are those so common about the Cape of *Good Hope*, or the following.

---

* With several sketches of the same, and an ample description, in a letter, *July* 14, 1770.

† *Niebuhr, Descr. Arabie,* 147.

C c 2

B. with

104. URSINE.

B. with a great head and long thick nose: short ears: crown covered with long upright hairs. The part of the head immediately above the forehead prominent, and terminating in a ridge. The whole body covered with long dusky hair, so that at first sight the animal appears like a young bear.

Body thick and strong: limbs short: tail half the length of the body; strait at the beginning, arched at the end: nails flat and round: buttocks of a bloody redness.

Is four feet high, even when sitting; and as tall as a middle-sized man, when erect.

Inhabits the *Cape of Good Hope*. Are very numerous, and go in troops in the mountains. When they see any one approach, they set up an universal and horrible cry for about a minute or two, and then conceal themselves in their fastnesses, and keep a profound silence. They hardly ever descend into the plains, unless it be to pillage the gardens, which lie at the foot of the mountains. It is said, that while they are plundering, they place centinels to guard against surprize; and that for greater expedition, they fling the fruit from one to another, in order to carry it off. They break the fruit into pieces, and cram it into the pouches nature hath furnished them with on each side of their cheeks, in order to eat it afterwards at leisure. The centinel, on sight of man, gives a yell; when the whole troop retreats in the most diverting manner, the young clinging to the backs of the parents *.

* *Kolben*, ii. 120. *La Caille*, 296.

When

When taken and confined, are tolerably tame; but very revengeful when provoked. They are ftrong enough to draw the ftrongeft man to them, notwithftanding he makes the moft powerful refiftance. They ufually lay hold of the ears, and will bite off one as clofe as if it was done with a razor.

This feems to be the fame with the *Mandrill*, defcribed by *Smith* in his voyage to *Guinea*, which he fays grows to a great fize, and that the body is as thick as á man's. Head very large: face covered with a white fkin: nofe always a running; and body cloathed with long black hair, like a bear.

Le Papion ou Babouin proprement dit. *de Buffon*, xiv. 133. tab. *Schreber*, 98. tab. vi.   105. MOTTLED.

B. with the nofe covered with a dufky red fkin. Hair on the head, neck, fhoulders, and breaft, very long; in other parts fhorter. Colors a mixture of tawny, black, and brown: feet dufky: buttocks naked, and hideous.

Tail, in the fpecimen defcribed by M. *de Buffon*, only feven inches long, it being mutilated: nails on the thumbs; on the toes, blunt and crooked claws.

Height, when fitting up, fometimes three or four feet: has all the deteftable manners of the former.   SIZE.

From the defect in the tail, it is difficult to determine the fpecies, or be certain whether it fhould be placed with thefe long-tailed baboons, or as a connecting link between them and the fhorter.

M. *de*

β. LITTLE.

M. *de Buffon* has deſcribed and engraven another, which he calls *le Petit Babouin*, differing only in ſize from the other, being a quarter leſs; but I fall into Mr. *Schreber*'s notion of its being only a young animal. See the former's account and figure, p. 147, tab. xiv. The latter's, p. 99, tab. vi. fig. 2.

106. LION-TAIL-ED.

Cercophithecus barbatus primus. *Cluſii exot.* 371. *Raii ſyn. quad.* 159. *Klein quad.* 89.
Simia veter. S. caudata barbata alba barba nigra. *Lin. ſyſt.* 36. *Briſſon quad.* 147.
Simia ſilenus. S. caudata barbata nigra, barba nigra prolixa. *Lin. ſyſt.* 36. *Briſſon quad.* 149.

Cercopithecus niger Ægyptiacus, *ibid.*
Simia Faunus. S caudata barbata, cauda apice floccoſa. *Lin. ſyſt.* 36.
Cercopithecus barbatus infra albus, barba incana mucronata, cauda in floccum deſinente. *Briſſon quad.* 144.
Le ſinge barbu noir. *Schreber,* 107, tab. xi. Mus. LEV.

B. with a long dog-like face, naked, and of a duſky color: a very large and full white or hoary beard: the beards of theſe females brown: large canine teeth: body covered with black hair: belly of a lighter color: nails flat: tail terminated with a tuft of hair like that of a lion: bulk of a middling-ſized dog.

Inhabits the *Eaſt Indies*, and the hotter parts of *Africa*.

One was ſhewn in *London* ſome years ago, exceſſively fierce and ill-natured: the tail not longer than the back, ending with a large tuft: beard reaching quite up the cheeks, as far as the eyes. This is certainly the *Ouanderou* of M. *de Buffon*, xiv. 169. *tab.* xviii. which he makes a ſort of Baboon, or Monkey with a ſhort tail; for he ſeems to have met with a ſpecimen mutilated in that part; and deſcribes it accordingly.

To theſe may be added the following more obſcure ſpecies.

δ. The

1. Lion tail'd Baboon ___ 106.

2. Tawny Monkey ___ 126.

δ. The little bearded men of *Barbot, voy. Guinea,* 212. and *Bof-man,* 242. are about two feet high, and are black as jet, with long white beards. The negroes fet a great value on the fkins of this fpecies, and fell them to one another at eighteen or twenty fhillings each. Of the fkins of thefe they make the caps for the *Tie-tie's,* or public Criers.

*** With tails longer than their bodies, or Monkies.

A. Thofe of the old world, or the continents of *Afia* and *Africa,* having within each lower jaw pouches for the reception of their food.

Buttocks (generally) naked.

Tails ftrait, not prehenfile.

M. with a great triangular white beard, fhort and pointed at the bottom; and on each fide of the ears extending in a winged fafhion far beyond them: face and hands purple: body black: tail much longer than the body, terminated with a dirty white tuft.

107. PURPLE-FACED.

Inhabit *Ceylon.* The figure taken from a drawing communicated to me by Mr. *Loten,* is probably the fame with thofe called by *Knox* * *Wanderows.* Thefe are very harmlefs; live in the woods, and feed on leaves and buds of trees; and when taken foon become tame.

There is a variety entirely white; but in form exactly like the others. Thefe are much fcarcer †.

* *Hift. Ceylon,* 25.   † The fame.

4

This

This is defcribed in the former edition, p. 109, β. as a variety; but on reconfideration, is here placed as a diftinct fpecies.

108. PALATINE.

La Palatine. *Schreber*, i. 124, *tab*. xxv.
La Palatine, ou Roloway. *Allamande*, 77. LEV. MUS.

M. with a triangular black face, bordered all round with white hair, which on the chin is divided into a long forked beard: back dufky: head, fides, and outfides of the arms and thighs, of the fame color, but each hair tipped with white: breaft, belly, and infide of the limbs, white, in the fubject fhewn in *Europe*, but in their native country orange, for they fade in our colder climate.

About a foot and a half high: the tail of the length of the body.

Inhabits *Guinea*; is called there *Roloway*: very full of frolic, and fond of the perfons it is acquainted with; averfe to others.

109. HARE-LIP-
PED.

Cercopithecus angolenfis major, macaquo. *Marcgrave, Brafil.* 227. *Raii fyn. quad.* 155. *Klein quad.* 89.
Cercopithecus cynocephalus, naribus bifidis elatis, natibus calvis. *Briffon quad.* 152. C. Cynoceph. ex virid. &c. 151.

S. Cynomolgus. S. caudata imberbis, naribus bifidis elatis, cauda arcuata, natibus calvis. *Lin. fyft.* 38. S. cynocephalus. *ibid.* Le Macaque. *De Buffon*, xiv. 190. *tab.* xiv. *Schreber*, 112.
Le Malbrouc. *Schreber*, 110. MUS. LEV.?

M. with the noftrils divided, like thofe of a hare: nofe thick, flat, and wrinkled: head large: eyes fmall: teeth very white: body thick, and clumfy: buttocks naked: tail long: color

color varies; fometimes like that of a wolf; but others are brown, tinged with yellow, or olive: belly and infide of the limbs of a light afh-color: the tail is rather fhorter than the body, and is alway carried arched.

Inhabits *Guinea* and *Angola*: is full of frolic, and ridiculous grimaces.

Le Malbrouck of M. *de Buffon,* xiv. 224. *tab.* xxix. fo much refembles this fpecies, that I place it here as a variety. That able Zoologift fufpected the fame; but feparates them, on account of fome trifling diftinctions, and the difference of country: this being a native of *India,* the other of *Africa:* but fince thofe very diftinctions may arife from the laft caufe, it feems better to unite them, than to multiply the fpecies, already fo numerous. A few years ago, one that feemed of this fpecies was fhewn in *London,* equal in fize to a fmall greyhound.

Cercopithecus barbatus Guineenfis, Exquima. *Marcgrave Brafil.* 227. *Raii fyn. quad.* 156.
Cercopithecus barbatus fufcus punctis albis infperfis barba alba. *Briffon quad.*

147. *No.* 23. 148. *No.* 24.
Simia Diana. S. caudata barbata, fronte barbaque faftigiata. *Lin. fyft.* 38.
L'Exquima. *De Buffon,* xv. 16.
La Diane. *Schreber,* 115. *tab.* xiv.

110. SPOTTED.

M. with a long white beard: color of the upper parts of the body reddifh, as if they had been finged, marked with white fpecks: the belly and chin whitifh: tail very long: is a fpecies of a middle fize.

Inhabits *Guinea* * and *Congo,* according to *Marcgrave:* the Con-

PLACE.

* *Purchas's Pilgrims,* ii. 955.

D d

gefe

gefe call it *Exquima.*   M. *de Buffon* denies it to be of that country : but, from the circumftance of the curl in its tail, in *Marcgrave*'s figure, and the defcription of fome voyagers, he fuppofes it to be a native of *South America.*

Linnæus defcribes his *S. Diana* fomewhat differently : he fays it is of the fize of a large cat; black, fpotted with white : hind part of the back ferruginous : face black : from the top of the nofe is a white line paffing over each eye to the ears, in an arched form : beard pointed; black above, white beneath; placed on a fattifh excrefcence : breaft and throat white : from the rump, crofs the thighs, a white line : tail long, ftrait, and black : ears, and feet, of the fame color : canine teeth, large.

III. LONG-NOSED.   **M.** with a very long flender nofe, covered with a flefh-colored naked fkin : hair on the head falling back; on the body and breaft long : color of the head, and upper part of the body and limbs, pale, ferruginous, mixed with black; of the breaft and belly light afh : tail very long.

SIZE: Height when fitting down about two feet : very good-natured. Defcribed from a drawing by Mr. *Paillou*, animal painter.   Place uncertain, probably *Africa.*   Its face very like that of a long-nofed dog.

I engrave another in the fame plate, under the title of the *Prude*, which poffibly may be related to the former.

M. with

1. Long nosed Monkey___. V.III.

2. ___ Prude Monkey ___. V.III.

M. with a black face: great canine teeth: great black naked ears: on the fide of the cheeks long hairs, of a pale yellow, falling backward towards the head: long hairs above each eye: throat and breaft of a yellowifh white: crown, upper part of the body, arms, and thighs, cinereous, mixed with yellow. On the lower part of the arms and legs, and on the tail, the cinereous predominates. Hair on the body coarfe. Tail the length of the body.

Size of a fox.

Inhabits *Guinea?* LEV. MUS.

112. YELLOWISH.

Simius Callitrichus. *Prof. Alp. Ægypt.* i. Simia fabæa. S. caudata imberbis flavicans, facie atra, cauda cinerea, natibus calvis. *Lin. fyft.* 38. *Edwards*, 215. Cercopithecus ex cinereo flavefcens, genis longis pilis albis obfita. *Briffon quad.*

145. et Cercobarbatus rufus facie nigra, cæfarie alba cincta. 149. Le Callitriche. *De Buffon*, xiv. 272. *tab.* xxxvii. *Schreber*, 122. *tab.* xviii. MUS. LEV.

113. GREEN.

M. with a black nofe: red flattifh face: the fides of it bounded by long yellow hairs, falling backwards like a muftachio, and almoft covering the ears, which are black, and like the human: head, limbs, and whole upper part of the body and tail, covered with foft hairs, of a yellowifh green color at their ends, cinereous at their roots: under fide of the body and tail, and inner fide of the limbs, of a filvery color: tail very long and flender: fize of a fmall cat.

Inhabits different parts of *Africa*: keep in great flocks, and

D d 2

live

MONKIES.

live in the woods : are scarce discernible when among the leaves, except by their breaking the boughs with their gambols, in which they are very agile and silent : even when shot at, do not make the lest noise; but will unite in company, knit their brows, and gnash their teeth, as if they meant to attack their enemy * : are very common in the *Cape Verd* islands : and also found in the *East Indies :* from whence Sir *A. Lever* had his specimen.

114. WHITE EYELID.

β Simia *Æthiops.* caudata imberbis, capillitio erecto lunalaque frontis albis. *Lin.syst.* 39. *Hasselquist itin?* 190. *Shaw, spec. Lin.* iv.

Le Mangabey. *De Buffon,* xiv. 244. *tab.* xxxii. xxxiii. *Schreber,* 128. *tab.* xx. xxi. Lev. Mus.

M. with a long, black, naked, and dog-like face : the upper eye-lids of a pure white : ears black, and like the human : no canine teeth : hairs on the sides of the face, beneath the cheeks, longer than the rest : tail long : color of the whole body tawny and black : flat nails on the thumbs and fore-fingers; blunt claws on the others : tail, hands, and feet black.

Shewn in *London* a few years ago : place uncertain : that described by M. *de Buffon* came from *Madagascar :* was very good-natured, went on all fours.

Le Mangabey a collier blanc †, is a variety, with the long hairs on the cheeks and round the neck white.

I have seen one at Mr. *Brook*'s, perhaps of this kind, with the crown of the head ferruginous : cheeks, under side of the neck, and belly, white : back, legs, and tail black.

* *Adanson's voy.* 316.

†. Of M. *de Buffon, tab.* xxxiii.

Cercopithecus

Cercopithecus alius Guineensis. *Marc-grave Brasil.* 228. *Raii syn. quad.* 156. S. cephus. S. caudata buccis barbatis, vertice flavescente, pedibus nigris, caudæ apice ferruginea. *Lin. syst.* 39. Cercopithecus nigricans, genis et auri-culis longis pilis ex albo flavicantibus obsitis, ore cærulescente. *Brisson quad.* 146. Le Moustac. *De Buffon,* xiv. 283. *tab.* xxxix. *Schreber,* 125. *tab.* xix. LEV. MUS.

M. with a short nose, of a dirty blueish color; beneath the nose a transverse stripe of white: edges of both lips, and space round the eyes, black: on the cheeks, before the ears, two large tufts of yellow hairs, like *mustaches:* ears round, and tufted with whitish hairs: the hair on the top of the head long and upright: round the mouth are some black hairs: the color of the hair on the head yellow, mixed with black: on the body and limbs, a mixture of red and ash-color: the part of the tail next the body of the same color; the rest yellowish: the under part of the body paler than the upper: the feet black: nails flat: its length one foot; that of the tail, eighteen inches.

Inhabits *Guinea.*

Simia nictitans. S. caudata imberbis nigra punctis pallidis aspersa, naso albo, pollice palmarum brevissimo, natibus tectis. *Lin. syst.* Cercopithecus Angolensis alius *. *Marc-grave Brasil.* 227. White Noses. *Purchas's Pilg.* ii. 955. LEV. MUS.

M. with a black flat face: the end of the nose of a snowy whiteness: irides yellow: hair on the head and body smooth, mottled with black and yellow: belly white: hands black: tail very long; upper side black, lower white.

Inhabits

Inhabits *Guinea* and *Angola :* is when tamed, after being taken young, very fportive and diverting : in a wild ftate avoids mankind : is very crafty, and has a very bad fmell.

The ape defcribed by Mr. *Schreber,* p. 126. tab. xix. B. agrees with this in the whitenefs of the nofe, but has a large white beard, which that which I faw wanted. He calls it, *Le Blanc Nez ;* and *Simia Petaurifta.*

117. TALAPOIN.   Le Talapoin. *De Buffon,* xiv. 287. *tab.* xl. *Schreber,* 124. *tab.* xvii. LEV. MUS.

M. with a fharp nofe, round head, large black naked ears : eyes, and end of the nofe, flefh-colored : hair on the cheeks very long, and reflected towards the ears : on the chin a fmall beard : the color of the whole upper part of the body, a mixture of dufky yellow and green : outfide of the limbs black ; infide whitifh : the lower part white tinged with yellow : the tail very long and flender ; above, of an olive and dufky color ; beneath, cinereous : the paws black : length, about one foot ; of the tail, one foot five inches.

Inhabits *India.*

118. NEGRO.   Middle-fized black-monkey. *Edw.* 311. *Schreber,* 131. *tab.* xxii. B. LEV. MUS.

M. with a round head : nofe a little fharp : face, of a tawny flefh-color, with a few black hairs : irides, a reddifh hazel : hair above the eyes long, uniting with the eye-brows ; that

on

on the temples partly covering the ears : breaft and belly of a fwarthy flefh-color, almoft naked : hair on the body, limbs, and tail, black, and pretty long : paws covered with a black foft fkin : fize of a large cat.

Inhabits *Guinea:* active, lively, entertaining, good-natured. In *Siam* is a large fpecies of black monkey, probably different from this.

S. aygula. S. caudata fubimberbis grifea, eminentia pilofa verticis reverfa longi-tudinali. *Lin. fyft.* 39. *Ofbeck's voy.* i.

151.
L'Aigrette. *De Buffon*, xiv. 190. *tab.* xxi. *Schreber*, 129. *tab.* xxii.

M. with a long face, and an upright fharp-pointed tuft of hair on the top of the head : hair on the forehead black : the color of the upper part of the body olivaceous ; of the lower cinereous : eye-brows large : beard very fmall : fize of a fmall cat.

Inhabit *Java:* fawn on men, on their own fpecies, and embrace each other ; play with dogs, if they have none of their own fpecies with them : if they fee a monkey of another kind, greet him with a thoufand grimaces : when a number of them fleep, they put their heads together : make a continual noife during night.

M. with a high, upright, rufty tuft on the crown : limbs and body ferruginous, mixed with dufky : belly, and infide of the legs, whitifh.

This fpecies is called by the *Malayes, Monèa,* from which is derived the *Englifh* name *Monkey.*

Le

121. RED.   Le Patas a bandeau noir. *De Buffon*, xiv. 208. *tab.* xxv.
                              Le Singe rouge. *Schreber*, 120. *tab.* xvi.

M. with a long nofe: eyes funk in the head: ears furnifhed with pretty long hairs: hairs on each fide of the face long: chin bearded: body flender: over each eye, from ear to ear, extends a black line: the upper part of the body of a moft beautiful and bright bay, almoft red, fo vivid as to appear painted: the lower parts afh-color, tinged with yellow: tail not fo long as the body, whofe length is about one foot fix inches.

M. *de Buffon* gives a variety of this fpecies, *tab.* xxvi, with a white band crofs the face, which he calls *Le Patas a bandeau blanc.*

Inhabits *Senegal:* is lefs active than the other kinds: very inquifitive: when boats are on their paffage on the river, will come in crowds to the extremities of the branches, and feem to admire them with vaft attention: at length, will become fo familiar, as to throw pieces of fticks at the crew: if fhot at, will raife hideous cries; fome will throw ftones, others void their excrements in their hands, and fling them among the paffengers *.

Barbot † mentions another fort of red monkey, called in *Guinea, Peafants,* becaufe of their ugly red hair and figure, and their natural ftink and naftinefs.

* *De Brue,* as quoted by M. *de Buffon.*   † *Defcr. Guinea,* 212.
5

122. CHINESE.

Rillow. *Knox's Ceylon*, 26.
Le Bonnet-Chinois. *De Buffon*, xiv. 190. *tab*. xxx. *Br. Muf. Schreber*, 132. *tab*.
xxiii. LEV. MUS.

M. with a long fmooth nofe, of a dufky color: hair on the
crown of the head long, lying flat, and parted like that of
a man: color, a pale cinereous brown, mixed with yellow: belly
whitifh.

In the LEVERIAN MUSEUM is a variety of a ferruginous color,
with a dufky face, and naked hands.

Inhabit *Ceylon*: keep in great troops: rob the gardens of fruit,
and fields of the corn: the natives are obliged to watch the whole
day; yet thefe monkies are fo bold, that, when drove from one
end of the field, they will immediately enter at the other, and carry
off with them as much as their mouth and arms can hold. *Bofman* *, fpeaking of the thefts of the monkies of *Guinea*, fays, that
they will take in each paw one or two ftalks of millet, as many
under their arms, and two or three in their mouth; and thus la-
den, hop away on their hind legs; but if purfued, fling away all,
except what is in their mouths, that it may not impede their
flight. They are very nice in their choice of the millet, examine
every ftalk, and if they do not like it, fling it away; fo this delicacy does more harm to the fields than their thievery. Of late
years a *Ruffian* tanner has difcovered that the fkins might be dreffed
and made into fhoes.

* *Voy. Guinea*, 243.

123. BONNETED.

M. with a dusky face: on the crown a circular bonnet, consisting of upright black hairs: on the sides of the cheeks the hairs are long: those and the body brown: legs and arms black.

Size of a small cat. LEV. MUS.

124. VARIED.

Κηβος? *Arist. hist. An.*
Monne? *Leo Afr.* 342.
Monichus. *Prosp. Alp. Ægypt.* i. 242.
La Mone. *De Buffon*, xiv. 258. *tab.* xxxvi.
  Schreber, 119. *tab.* xv.

Cercopithecus pilis ex nigro et rufo variegatis vestibus, pedibus nigris, cauda cinerea. Le singe varié. *Brisson quad.* 141. LEV. MUS.

M. with a short, black, thick nose: orbits and mouth of a dirty flesh-color: hair on the sides of the face, and under the throat, long, of a whitish color, tinged with yellow: on the forehead, grey: above the eyes, from ear to ear, a black line: the upper part of the body dusky and tawny: the breast, belly, and inside of the limbs white: outside of the thighs, and arms, black: hands and feet black and naked: the tail of a cinereous brown: length, about a foot and a half; the tail above two.

Inhabits *Barbary, Æthiopia*, and other parts of *Africa*: is the kind which gives the name of Monkey to the whole tribe, from the *African* word *Monne*; or rather its corruption, *Monichus*. M. *de Buffon* supposes it to be the Κηβος of *Aristotle*: but the Philosopher says no more, than that the *Cebi* are apes furnished with tails.

Of this kind is the *Cercopithecus Guineensis alius* of *Marcgrave Brasil.*

Brasil. 228. *Brisson quad.* 139. which the first describes as being of the color of the back of a hare.

Le Douc. *de Busson,* xiv. 2c8. *tab.* xli.
Cercopithecus cinereus, genis longis pilis ex albo flavicantibus obsitis, torque ex castaneo purpurascente. Le grand singe de la *Cochin-chine. Brisson quad.* 146. *Schreber,* 137. tab. xxiv.

125. COCHIN-CHINA.

M. with a short flattish face, bounded on each side by long hairs of a yellowish color: on the neck a collar of purplish brown: the lower part of the arms, and tail, are white: the upper part of the arms, and thighs, black: legs and knees of a chesnut-color: the back, belly, and sides, grey, tinged with yellow: above the root of the tail is a spot of white, which extends beneath as far as the lower part of the belly and part of the thighs: the feet black: the buttocks * covered with hair: is a very large species, about four feet long, from the nose to the tail; but the tail not so long.

Inhabits *Cochin-China* and *Madagascar* †: lives on beans: often walks on its hind feet.

M. with a face a little produced: that and the ears flesh-colored: nose flattish: long canine teeth in the lower jaw: hair on the upper part of the body pale tawny, cinereous at the

126. TAWNY.

---

* All the species of apes of *Asia* and *Africa,* except this and No. 64, 70, and 87, have their buttocks naked.

† Where it is called *Sifac, Flacourt hist. Madag.* 153.

roots :

roots: hind part of the back orange: legs cinereous: belly white; fize of a cat: tail fhorter than the body.

Inhabits *India*. From one in Mr. *Brookes's* exhibition. Very ill-natured.

M. *Paillou* communicated to me a variety of this fpecies, with a black face, and long black hairs on the cheeks: body of a dull pale green: limbs grey: tail dufky.

127. GOAT.

M. with a blue naked face ribbed obliquely: a long beard, like that of a goat: whole body and limbs of a deep brown color: tail long. Defcribed from a drawing in the *Britifh Mufeum*, by *Kikius*, an excellent painter of animals.

128. FULL-BOT-TOM.

M. with a fhort, black, and naked face: fmall head; that and the fhoulders covered with long, coarfe, flowing hairs, like a full-bottomed perriwig; of a dirty yellowifh color, mixed with black: body, arms, and legs, of a fine gloffy blacknefs, covered with fhort hairs.

Hands naked, furnifhed with only four fingers: on each foot five very long flender toes.

Tail very long; of a fnowy whitenefs; with very long hairs at the end, forming a tuft: body and limbs very flender: length above three feet.

Inhabits the forefts of *Sierra Leone*, in *Guinea*; is called there, *Bey* or *King Monkey*: the negroes hold its fkin in high eftima-tion, and ufe it for pouches, and for coverings to their guns.

M. with

Full bottom Monkey ___. N.º 128.

M. with a black crown: back of a deep bay color: outside of the limbs black: cheeks, under part of the body, and legs, of a very bright bay.

Only four fingers on the hands; on the feet five long toes.

Tail very long, flender, and black.

Body and limbs very flender and meagre.

Inhabits *Sierra Leone*, and brought over by Mr. *Smeathman*, who prefented this and the former to the LEVERIAN MUSEUM.

129. BAY.

Simia apella. *Lin. fyft.* 42. *Schreber*, tab. xxviii.

M. with a flat face: long hairs on the forehead and cheeks: upper part of the body and limbs of a tawny brown; belly cinereous: tail fhorter than the body, annulated with a darker and lighter brown: hands naked and black. From a drawing in the *Britifh Mufeum*.

130. ANNULAT-ED.

Cercopithecus *Luzonicus* minimus, *Magu* vel *Root* Indorum. *Pet. Gaz.* 21. *tab.* xiii.

Simia fyrichta. S. caudata imberbis ore ciliifque vibriffatis. *Lin. fyft.* 44. *Schreber,* 152. *tab.* xxxi.

131. PHILIPPINE.

M. with its mouth and eye-brows befet with long hairs: an obfcure fpecies, mentioned only by *Petiver*; faid to come from the *Philippine* ifles.

8

B. Monkies

B. Monkies of the new world, or the continent of *America*, having neither pouches in their jaws, nor naked buttocks.
Tails of many prehensile, and naked on the under side, for a certain space next their end.

α. With prehensile tails *.

132. PREACHER. Guariba. *Marcgrave Brasil*, 226. Rai *syn. quad.* 153.
Aquiqui. *De Laet*, 486. *Grew's Museum*, 11.
Howling Baboons, Guareba. *Bancroft's Guiana*, 133.
Simia Beelzebub. S. caudata barbata nigra, cauda prehensili extremo pedibusque fuscis. *Lin. syst.* 37.
Cercopithecus niger, pedibus fuscis. *Brisson quad.* 137.
L'Ouarine. *Schreber*, 137. *de Buffon*, xv. 5.

M. with black shining eyes: short round ears: a round beard under the chin and throat: hairs on the body of a shining black; long, yet lie so close on each other that the animal appears quite smooth: the feet and end of the tail brown; tail very long, and always twisted at the end: size of a fox.

Inhabits the woods of *Brasil* and *Guiana* in vast numbers; and makes a most dreadful howling: sometimes one mounts on a higher branch, the rest seat themselves beneath: the first begins as if it was to harangue, and sets up so loud and sharp a howl as may be heard a vast way; a person at a distance would think that a hundred joined in the cry; after a certain space, he gives a signal with his hand, when the whole assembly joins in chorus; but

* These M. *de Buffon* calls *Sapajous*.

on

on another fignal, is filent, and the orator finifhes his addrefs *: their clamor is the moft difagreeable and tremendous that can be conceived, owing to a hollow and hard bone placed in the throat, which the *Englifh* call the *throttle*-bone †. Thefe monkies are very fierce, untameable, and bite dreadfully.

α. ROYAL. Cercopithecus barbatus maximus, ferruginofus, ftertorofus. *Alaoiita, finge rouge. Barrere, France Æquin.* 150.	Simia feniculus. S. caudata barbata rufa, cauda prehenfili. *Lin. fyft.* 37.
Cercopithecus barbatus-faturatè rufus. *Briffon quad.* 147.	Arabata. *Gumilla Orenoque,* ii. 8. *Bancroft Guiana,* 135.
	L'Allouatte, *de Buffon,* xv. 5. *Schreber,* 138.

A variety of a ferruginous or reddifh bay color, which the *Indians* ‡ call the king of the monkies: is large, and as noify as the former. The natives eat this fpecies, and feveral other forts of monkies, but are particularly fond of this; *Europeans* will alfo eat it, efpecially in thofe parts of *America* where food is fcarce: when it is fcalded, in order to get off the hair, it looks very white, and has a refemblance fhocking to humanity, that of a child of two or three years old, when crying ‖.

* A fingular account, yet related by *Marcgrave* and feveral other writers. *Marcgrave* is a writer of the firft authority, and a moft able naturalift, long refident in the *Brafils,* and fpeaks from his own knowledge.

† *Grew's Rarities,* 11.

‡ *De Laet.* 486.

‖ *Ulloa's voy.* i. 113. *Des Marchais,* iii. 311, fays, they are excellent eating, and that a *foupe aux finges* will be found as good as any other, as foon as you have conquered the averfion to the *Bouilli* of their heads, which look very like thofe of little children.

Cercopithecus

133. FOUR-FIN-
GERED.

Cercopithecus major niger, faciem huma-
nam referens. Quouata. *Barrere France
Æquin.* 150.
Quato. *Bancroft Guiana,* 131.
Cercopithecus in pedibus anterioribus
pollice carens cauda inferius apicem
versus pilis deftituta. Le Belzebut.
Briffon quad. 150.

Simia Panifcus. S. caudata imberbis atra,
cauda prehensili, ad apicem subtus nuda. *Lin. fyft.* 37.
Le Coaita. *de Buffon,* xv. 16. *Schreber,*
140. tab. xxvi.
Spider Monkey *Edw. Gleanings,* iii. 222.
br. Mu. Lev. Mus.

M. with a long flat face, of a swarthy flesh-color: eyes funk
in the head: ears like the human: limbs of a great
length, and uncommonly slender: hair black, long, and rough:
only four fingers on the hands, being quite deftitute of a thumb:
five toes on the feet: nails flat: tail long, and naked below near
the end: body slender: about a foot and a half long: tail near
two feet, fo prehensile as to ferve every purpofe of a hand.

Inhabits the neighborhood of *Carthagena, Guiana, Brafil,*
and * *Peru:* affociate in vaft herds: fcarce ever are feen on the
ground. *Dampier†* defcribes their gambols in a lively manner:
' There was,' fays he, ' a great company, dancing from tree to
' tree over my head, chattering and making a terrible noife,
' and a great many grim faces and antic geftures; fome broke
' down dry fticks and flung at me, others fcattered their urine
' and dung about my ears at laft one, bigger than the reft, came
' to a fmall limb juft over my head, and leaping directly at me,
' made me leap back, but the monkey caught hold of the bough
' with the tip of his tail, and there continued fwinging to and
' fro, making mouths at me. The females with their young ones

* *De Buffon,* xv. 21.   † *Voy.* ii. 60.

' are

' are much troubled to leap after the males, for they have com-
' monly two, one fhe carries under her arm, the other fits on her
' back, and claps its two fore paws about her neck. Are very
' fullen when taken; and very hard to be got when fhot, for
' they will cling with their tail or feet to a bough, as long as
' any life remains; when I have fhot at one, and broke a leg
' or arm, I have pitied the poor creature, to fee it look and handle
' the broken limb, and turn it from fide to fide.'

They are the moft active of monkies, and quite enliven the
forefts of *America:* in order to pafs from top to top of lofty trees,
whofe branches are too diftant for a leap, they will form a chain,
by hanging down, linked to each other by their tails, and fwinging in that manner till the loweft catches hold of a bough of the
next tree, and draws up the reft*, and fometimes they pafs †
rivers by the fame expedient.

Are fometimes brought to *Europe:* are very tender, and feldom live long in our climate: Mr. *Brookes* had one or two,
which, as long as they continued in health, were fo active, and
played fuch tricks, as to confirm the account of voyagers.

Simia trepida. S. caudata imberbis, capillitio arrecto, manibus pedibuf-que cæruleis, cauda prehenfili vil-lofa. *Lin. fyft.* 39.
Singe, &c. *Schreber,* 147. tab. xxvii.

Bufh-tailed Monkey, *Edw.* 312.
Simiolus Ceylonicus. *Seb. Muf.* i. 77. *tab.* 48. *Br. Muf.*
Le Sajou. *de Buffon,* xv. 37. tab. iv. v.

134. FEARFUL.

M. with a round head; and fhort flefh-colored face, with a lit-tle down on it: hair on the forehead more or lefs high and erect in different fubjects: top of the head black or dufky,

* *Wafer's voy.* in *Dampier,* iii. 330. † *Ulloa,* i. 113.

hair on it pretty long: hind part of the neck, and middle of the back, covered with long dufky hairs; reft of the back and the limbs of a reddifh brown: hair on the breaft and belly very thin: hands and feet covered with a black fkin: on the toes flat nails: tail longer than the head and body, and often carried over the fhoulders; the hair on it very long, of a deep brown color, and appears very bufhy from beginning to end.

Inhabits *Guiana*, not *Ceylon*, as *Seba* afferts: is a lively fpecies; but capricious in its affections in a ftate of captivity, having a great fondnefs for fome perfons, and as great a hatred to others.

135. CAPUCIN. Simia capucina. S. caudata imberbis fufca, cauda prehenfili hirfuta, pileo artubufque nigris, natibus tectis. *Lin.*   *fyft.* 42. *Muf. Ad. Fred.* tab. ii. Le fai. *Schreber,* 147. tab. xxix. *de Buffon,* xv. 51. tab. viii. LEV. MUS.

M. with a round head: face flat and flefh-colored, encircled with upright whitifh hairs: breaft covered with long fhaggy pale yellow hair: head black; body and tail of a deep brown, or dufky: tail very long, and thickly cloathed: on the toes are crooked claws, not flat nails as on thofe of the former. I confefs my inattention to that circumftance in my former edition, which made me confound this and the laft fpecies.

Inhabits *South America*.   5

Cercopithecus

Cercopithecus *Brasiliensis* secundus *Clusii* exot. 372.

Cay? *De Laet*, 486. *Raii syn. quad.* 155.

Cercopithecus totus niger. *Brisson quad.* 139.

Le Sai—Le Sai a gorge blanc, *de Buffon*, xv. 51. tab. viii. ix.

Schreber, 147. tab. xxviii.

Simia apella. *Lin. syst.* 42. *Mus. Ad. Fred.* tab. i.

136. WEEPER.

M. with a round and flat face, of a reddish brown color, very deformed: the hair on the head, and upper part of the body, black, tinged with brown; beneath, and on the limbs, tinged with red: tail black, and much longer than the head and body: the young excessively deformed; their hair very long, and thinly dispersed: on each toe a flat nail. In the *British Museum* are specimens of old and young. *M. de Buffon* has a variety with a white throat.

Inhabits *Surinam* and *Brasil*: appear as if they were always weeping *: of a melancholy disposition; but very full of imitating what they see done. These probably are the monkies *Dampier* saw in the *Bay* of *All Saints*, which he says are very ugly, and smell strongly of musk †: keep in large companies; and make a great chattering, especially in stormy weather: reside much on a species of tree, which bears a podded fruit, which they feed on ‡.

The figure in *Mus. Ad. Fred.* has much too cheerful a countenance.

* *Froger's voy.* 116. † *Dampier's voy.* iii. 53. ‡ *De Laet,* 486.

Caitaia.

137. ORANGE.

Caitaia. *Marcgrave Brasil.* 227. *Raii syn. quad.* 175.
Cercopithecus pilis ex fusco, flavescente, et candicante variegatis vestitus, pedibus ex flavo rufescentibus. *Brisson quad.* 140.
Cercopithecus ex albo flavescens moschum redolens. *Brisson,* 139.
Cercopithecus minor luteus; Le Sapajou jaune. *Barrere France Æquin.* 151.
Simia-sciurea. S. caudata imberbis, occipite prominulo, unguibus quatuor plantarum subulatis, natibus tectis. *Lin. syst.* 43.
Le Saimiri. *de Buffon,* xv. *Mus.*
Schreber, 148. tab. xxx. LEV. MUS.

M. with a round head: nose a little pointed: tip of the nose, and space round the mouth, marked with black, of a circular form: orbits flesh-colored: ears hairy: hair on the body short, woolly, and fine, of a yellow and brown color; but in its native country, when in perfection, of a brilliant gold * color: the feet orange: nails of the hands flat: of the feet, like claws: tail very long; less useful for prehensile purposes than that of the rest: body of the size of a squirrel.

Inhabits *Brasil* and *Guiana:* when provoked, screams: is a very tender animal: seldom brought here alive: smells of musk †. The *Simia Morta* of *Linnæus,* 43; and *Cercopithecus cauda murina* of *Brisson,* 143; engraved in *Seba, tab.* 48. under the name of *Simiolus Ceylonicus,* is only the fœtus of some monkey: probably, as *Linnæus* conjectures, of this species.

* *Froger's voy.* 116.

† Some of the *African* monkies have also a strong smell of musk. A *Bezoar* is sometimes found in certain species.

Cercopithecus

Cercopithecus ex nigro et fuſco variegatus, faſciculis duobus pilorum capitis
corniculorum æmulis. Le Sapajou cornu. *Briſſon quad.* 138.
Simia Fatuellus. *Lin. ſyſt.* 42. Lev.
Mus.

M. with two black tufts of hair like horns on the top of the
head: eyes bright; of a duſky color: ears like the human: face, ſides, belly, and fore legs, reddiſh brown: upper part
of the arms, neck, and upper part of the back, yellowiſh: top
of the head, lower part of the back, hind legs, and all the feet,
black: tail prehenſile, covered with ſhort bright hair: body fourteen inches long, tail fifteen.

Inhabits *America.* A moſt deformed ſpecies.

M. with a ſhort noſe: black face: hair on each ſide long:
back and ſides orange and black, intimately mixed:
belly white: outſide of the legs black; inſide aſh-colored: tail of
a duſky aſh: its length twenty inches; that of the body eighteen.

In poſſeſſion of the late *Richard Morris,* Eſq; of the Navy-
Office: brought from *Antigua:* but its native place uncertain:
very good-natured, lively, and full of tricks: frequently hung
by its tail.

b. with

b. with ftrait tails, not prehenfile *

140. FOX-TAIL-
ED.

Cagui major. *Marcg Brafil.* 227.
Cercopithecus pilis nigris, apice albido,
veftitus, cauda pilis longiffimis nigris
obfita. *Briffon quad.* 138. C. pilis cine-
refcentibus nigro mixtis, cauda rufa.
Briffon, 141.
Simia Pithecia. S. caudata imberbis, vellere nigro apice albo, cauda nigra vil-
lofiffima. *Lin. fyft.* 40.
Le Saki. *De Buffon,* xv. 88. *tab.* xii. *Schreber,* 153. *tab.* xxxii.
Saccawinkee. *Bancroft Guiana,* 135. *Br.
Muf.* Mus. Lev.

M. with a fwarthy face, covered with fhort white down: forehead and fides of the face with whitifh, and pretty long
hair: body with long dufky brown hairs, white or yellowifh at
their tips: hair on the tail very long and bufhy; fometimes
black, fometimes reddifh: belly and lower part of the limbs a
reddifh white: length from nofe to tail near a foot and a half:
tail longer, and like that of a fox: hands and feet black, with
claws inftead of nails.

Inhabits *Guiana.*

* Diftinguifhed from thofe with prehenfile tails, by M. *de Buffon,* by the name
of *Sagouins*; which, as well as *Sapajous,* are *American* names for certain kinds of
monkies.

Cercopithecus

Cercopithecus minimus niger Leontoce-
  phalus, auribus elephantinis. *Barrere*
  France Æquin. 151.
Simia midas. S. caudata imberbis, labio
  fuperiore fiffo, auribus quadratis nudis,
  unguibus fubulatis, pedibus croceis.

Lin. fyft. 42.
Le Tamarin. *De Buffon,* xv. 92. *tab.* xiii.
  Schreber, 160. *tab.* xxxvii.
Little black monkey. *Edw.* 196. *Br. Muf.*
Lev. Mus.

141. GREAT.
EARED.

M. with a round head, fwarthy, flefh-colored, naked face: upper lip a little divided: ears very large, erect, naked, and almoft fquare: hair on the forehead upright and long; on the body foft, but fhaggy: the head, whole body, and upper part of the limbs, black, except the lower part of the back, which was tinged with yellow: hands and feet covered with orange-colored hairs, very fine and fmooth: nails long and crooked: tail black, and twice the length of the body: teeth very white.

Size of a fquirrel.

Inhabits the hotter parts of *South America,* and the ifle of *Gorgona,* fouth of *Panama,* in the *South Sea.* There are, fays *Dampier,* a great many little black monkies: at low water, they come to the fea-fide to take mufcles and perriwinkles, which they dig out of the fhells with their claws *.

PLACE.

* Voy. i. 173.

Cagui

142. STRIATED.

Cagui minor. *Marcgrave Brafil*, Cercopithecus *Brafilianus* tertius Sagouin.
Clufii Exot. 372. G*efner quad.* 869.
Raii fyn. quad. 154. *Kl in quad.* 87. *tab.*
iii. *Ludolph. Com. Æthiop.* 58.
Cercopithecus tæniis tranfverfis alternatim fufcis et e cinereo albis variegatus,
auriculis pilis albis circumdatis. *Briffon quad.* 143.
Simia Iacchus. S. caudata auribus villofis

patulis, cauda hirfutiffima curvata, un-
guibus fubulatis ; pollicum rotundatis.
Lin fyft. 40.
L'Ouiftiti. *De Buffon*, xv. 96. *tab* xiv.
Sanglin, or Cagui minor. *Edw.* 218. *Ph.
Tr. abridg.* 1751, *p.* 146. *tab.* vii. *Br.
Muf.*
Le Sagoin. *Schreber*, 154. *tab.* xxxiii.
MUS. LEV.

M. with a very round head: about the ears two very long full
tufts of white hairs ftanding out on each fide: irides reddifh: face of a fwarthy flefh-color: ears like the human: head
black: body afh-colored, reddifh, and dufky; the laft forms ftri-
ated bars crofs the body: tail full of hair, annulated with afh-
color and black: body feven inches long: tail near eleven: hands
and feet covered with fhort hairs: fingers like thofe of a fquirrel:
nails, or rather claws, fharp.

Inhabits *Brafil:* feeds on vegetables; will alfo eat fifh*: makes
a weak noife: very reftlefs: often brought over to *Europe.*

* *Edw. Gleanings*, p. 17.

3

Cercopithecus

Silky Monkey.—N.º 143.

Cercopithecus minor dilutè olivaceus, parvo capite, Acarima *a Cayenne. Barrere, France Æquin.* 151.

Cercopithecus ex albo flavicans, faciei circumferentia, faturaté rufa. De petit finge Lion. *Briſſon quad.* 142.

Simia Rofalia. S. caudata imberbis, ca-

pite pilofo, faciei circumferentia pedibufque rubris, unguibus fabulatis. *Lin. fyſt.* 41.

Le Marikina. *De Buffon*, xv. 108. *tab.* xvi. *Schreber*, 158. *tab.* xxxv. Lev. Mus.

143. Silky.

M. with a flat face, of a dull purple color: ears round and naked: on the fides of the face the hairs very long, turning backwards, of a bright bay color; fometimes yellow, and the former only in patches: the hair on the body long, very fine, filky, gloffy, and of a light but bright yellow: hands and feet naked, and of a dull purple color: claws, inftead of nails, to each finger: length of the head and body ten inches: tail thirteen and a half; a little bufhy at the end.

Inhabits *Guiana*; is very gentle, and lively.

Pinche. *Condamine's voy.* 83.

Simia Œdipus. S. caudata imberbis, capillo dependente, cauda rubra, unguibus fubulatis. *Lin. fyſt.* 41.

Cercopithecus pilis ex fufco et rufo veftitus, facie ultra auriculas ufque ni

gra et nuda, vertice longis pilis obfita. *Briſſon quad.* 150.

Le Pinche. *De Buffon*, xv. 114. *tab.* xvii. *Schreber*, 156. *tab.* xxxiv.

Little Lion Monkey. *Edw.* 195.

144. Red-tailed.

M. with a round head and black pointed face: ears round and dufky: hair on the head white, long, and fpreading over the fhoulders: fhoulders and back covered with long and loofe brown hairs: rump and half the tail deep orange-colored, almoft red; the remaining part black: throat black: breaft, belly, and legs, white; infides of the hands and feet black: claws crooked and fharp: length of the head and body eight inches; tail above twice as long.

Inhabits *Guiana*, *Brafil*, and the banks of the river of *Amazons*, whofe woods fwarm with numberlefs fpecies: is agile and lively, and has a foft whiftling note. Often marches with its tail over its back, appearing like a little lion.

145. FAIR.
A Sagoin, &c. *Condamine's voy.* 83.
Cercopithecus ex cinereo albus argenteus, facie auriculifque rubris fplendentibus, cauda caftanei coloris. *Brif.*
fon quad. 142.
Le Mico. *De Buffon*, xv. 121. *tab.* xviii.
Schreber, 159. *tab.* xxxvi.

M. with a fmall round head: face and ears of the moft lively vermilion color: body covered with moft beautiful long hairs, of a bright and filvery whitenefs, of matchlefs elegance: tail of a fhining dark chefnut: head and body eight inches long; tail twelve.

Inhabits the banks of the *Amazons*, difcovered by M. *de Condamine.*

Six

Tail-less Maucauco____ N°.116

Tail less . Maucauco ─── N.º 146.

Six cutting teeth, and two canine teeth in each jaw.
Sharp-pointed fox-like vifage.
Feet formed like hands, like the apes.

Animal elegantiffimum *Robinfoni. Raii
fyn. quad.* 161.
Cercopithecus *Ceylonicus,* feu Tardigradus dictus, major. *Seb. Muf. tab.* xlvii.
Klein quad. 86.

Lemur tardigradus. L. ecaudatus. *Lin.
fyft.* 44. *Shaw, Spec. Lin.* v.
Simia unguibus indicis pedum pofteri-
orum longis, incurvis, et acutis. *Briffon quad.* 134. LEV. MUS.

146. TAIL-LESS.

M. with a fmall head; fharp-pointed nofe: orbits furrounded
with a black circle, fpace between them white: from the
top of the head along the middle of the back, to the rump, a
dark ferruginous line, which on the forehead is bifurcated: ears
fmall: body covered with fhort, foft, and filky afh colored and
reddifh fur: toes naked: nails flat: thofe of the inner toe on
each hind foot long, crooked, and fharp: length from the nofe to
the rump fixteen inches.

Inhabits *Ceylon* and *Bengal*; lives in the woods, and feeds on
fruits: is fond of eggs, and will greedily devour fmall birds: has
the inactivity of the Sloth *, creeps flowly along the ground †: is
very tenacious of its hold, and makes a plaintive noife.

The inhabitants of *Bengal* call this animal *Chirmundi Billi,* or
Bafhful Billy. It fleeps, as I have feen one do in *London* in this
year, holding faft the wires of the cage with its claws. It makes
a plaintive noife, *Ai, Ai.* Its tongue is rough.

* *Vide* that article: this animal, notwithftanding its manners, cannot be ranked with the *Sloth,* having both cutting and canine teeth.

† I doubt not but the candor of Mr. *Schreber* will induce him to rectify his
mifreprefentations of this paffage.

G g 2   Sonnerat

Sonnerat voy. II. 142. tab. lxxvii.

M. with a produced dog-like vifage: fhort ears, briefly tufted: hair filky and thick: face and lower parts greyifh: rump white: whole upper part of the neck and body black: nails flat, but pointed at the ends: no tail.

MANNERS.		The largeft of the genus being three feet and a half high its note is that like a child's crying. Is a very gentle animal: when taken young, it is trained for the chace as dogs are. Inhabits *Madagafcar*, where it is called *Indri*, or *Man of the Wood*.

148. LORIS.		Animalculum cynocephalum, ceilonicum,		Le Loris. *De Buffon*, xiii. 210. *tab.*
			Tardigradum dictum, fimii fpecies.			xxx. *Schreber*, 162. *tab*. xxxviii.
			Seb. Muf. i. 55. *tab*. xxxv.				LEV. MUS.

M. with a produced dog-like vifage: forehead high above the nofe: ears large, thin, and rounded: body flender and weak: limbs very long and flender: thumb on each foot more diftinct, and feparate from the toes; on that, and the three out-moft toes, are flat nails: on the interior toe of every foot a crooked claw: no tail: the hair on the body univerfally fhort, and delicately foft: the color on the upper part tawny; beneath whitifh: fpace round the eyes dufky: on the head is a dart-fhaped fpot, with the end pointing to the interval between the eyes.

MANNERS.		Length from the tip of the nofe to the anus only eight inches.

It differs totally in form and in nature from the preceding. Notwithftanding the epithet of *Sloth* given in *Seba*, it is very
active,

active, afcends trees moſt nimbly; has the actions of an ape. If we credit *Seba*, the male climbs the trees, and taſtes the fruits before it preſents them to its mate.

Macaſſar fox. *Nieuhoff voy.* 361. Chitote, *Barbot.* 560.
Vary (1). *Flacourt, hiſt. Madag.* 153.
Simia-ſciurus lanuginoſus fuſcus, *Petiv. Gaz. tab.* xvii.
The Mongooz. *Edw.* 216.
Proſimia fuſca. Pr. fuſca, naſo pedibuſque albis. Pr. fuſca, rufo admixto,

facie nigra, pedibus fulvis. *Briſſon quad.* 156, 157.
Lemur Mongooz. L. caudatus griſeus, cauda unicolore. *Lin. ſyſt.* 44.
Le Mongooz. *De Buffon,* xiii. 174. *tab.* xxvi. *Schreber,* 166. *tab.* xxxix. LEV. MUS.

149. WOOLLY.

M. with orange-colored irides: ſhort rounded ears: end of the noſe black: eyes lodged in a circle of black: the ſpace between them of the ſame color: reſt of the noſe and lower ſides of the cheeks white: when in full health, the whole upper part of the body covered with long, ſoft, and thick fur, a little curled or waved, of a deep browniſh aſh-color: tail very long, covered with the ſame ſort of hair, and of the ſame color: breaſt and belly white: hands and feet naked, and duſky: nails flat, except that of the inner toe of the hind feet: ſize of a cat: varies, ſometimes with white or yellow paws, and with a face wholly brown.

Inhabits *Madagaſcar*, and the adjacent iſles: ſleeps on trees: turns its tail over its head to protect it from rain *: lives on fruits: is very ſportive and good-natured: very tender: found as far as *Celebes* or *Macaſſar*. This is the ſpecies M. *Sonnerat* calls

* *Cauche, voy. Madagaſcar,* 53.

3

Maquis

Maquis a bourres, vol. ii. p. 143; but his figure is not by any means accurate. *Linnæus* confounds this with Mr. *Edwards's* black maucauco, our 151st.

150. RING-TAIL.

Vari. *Flacourt, hift. Madag.* 153.
Mocawk. *Grofe's voy.* 41.
Maucauco. *Edw.* 197.
Profimia cinerea, caudâ cinctâ annulis alternatim albis et nigris. *Briffon quad.* 157. *Shaw, Spec. Lin.* vi.

Lemur Catta. L. caudatus, cauda albo nigroque annulata. *Lin. fyft.* 45. *Ofbeck's voy.* ii. 168.
Le Mococo. *De Buffon,* xiii. 173. *tab.* xxii. *Schreber,* 172. *tab.* xli. LEV. Mus.

M. with the end of the nofe black: ears erect: white face: black circles round the orbits: hair on the top of the head and hind part, deep afh-color: back and fides reddifh afh-color: outfides of the limbs paler: belly and infide of the limbs white: all its hair very foft, clofe, and fine, erect like the pile of velvet: tail twice the length of the body; is marked with numbers of regular rings of black and white; and when fitting, is twifted round the body, and brought over its head: nails flat, particularly thofe of the thumbs of the hind feet: infide of the hands and feet black: fize of a cat.

Inhabits *Madagafcar* and the neighboring ifles: is very good-natured, has all the life of a monkey, without its mifchievous difpofition: is very cleanly: its cry weak: in a wild ftate, goes in troops of thirty or forty: is eafily tamed when taken young: according to *Flacourt,* fometimes found white; *Cauche* in his voyage to *Madagafcar** alfo fpeaks of a white kind, which he fays grunts like fwine, and is called there *Amboimenes.*

* P. 53.

5

Vari,

Vari, ou Varicoſſi. *Flacourt, hiſt. Madag.*   *Schreber,* 171. *tab.* xl.
153. *Cauche, voy.* 53.   Lemur caudatus niger, collari barbato
Black Maucauco. *Edw.* 217.   *Lin. ſyſt.* 44.
Le Vari. *De Buffon,* xiii. 174. *tab.* xvii.

151. RUFFED.

M. with orange-colored irides: long hair round the ſides of the head, ſtanding out like a ruff: tail long: the color of the whole animal black, but not always, being ſometimes white, ſpotted with black; but the feet black: rather larger than the laſt.

Inhabits *Madagaſcar:* very fierce in a wild ſtate; and makes ſo violent a noiſe in the woods, that it is eaſy to miſtake the noiſe of two for that of a hundred: when tamed are very gentle and good natured. The hind thighs and legs of theſe three ſpecies are very long, which makes their pace ſideling, and bounding.

Le Tarſier. *de Buffon,* xiii. 87. tab. ix. LEV. MUS.

152. TARSIER.

M. with a pointed viſage; ſlender noſe, bilobated at the end: eyes large and prominent: ears erect, broad, naked, ſemi-tranſparent; an inch and a half long: between them, on the top of the head, is a tuft of long hairs: on each ſide of the noſe, and on the upper eye-brow, are long hairs.

In each jaw are two cutting and two canine teeth; which form an exception in this genus.

Four long ſlender toes, and a diſtinct thumb, on each foot: the lower part of each tuberous: the claws ſharp-pointed; but (except on the two interior toes of the hind feet) are attached to

the

the fkin: the thumbs of the hind feet are broad, and greatly dilated at their ends: hairs on the legs and feet fhort, white, and thin; tail almoft naked.: the greater part round and fcaly, like that of a rat; but grows hairy towards the end, which is tufted.

The penis pendulous; fcrotum and tefticles of a vaft fize, in proportion to the animal: hair foft, but not curled: of an afh-color, mixed with tawny.

Length from nofe to tail near fix inches; to the hind toes eleven and a half, the hind legs, like thofe of the *jerboa*, being of a great length: the tail nine inches and a half long.   Defcribed from two fine fpecimens in the cabinet of Doctor *Hunter*.

Inhabits the remoteft iflands of *India*, efpecially *Amboina*.   Is called by the *Macaffars*, *Podje* *.

153. BICOLOR.   *Miller's plates*; tab. xiii. Lemur bicolor, *Gm. Lin.* 44.

M. with a large white heart-fhaped fpot between the ears, pointing downwards: face, nofe, back, and fides, almoft as low as the belly, black: breaft, fhoulders, legs, reft of the fides, and belly, white: tail much longer than the body, thickeft at the end, black: limbs ftrong: toes long and flender: nails long, ftrait, and very flender: feet an exception to the genus.   Inhabits *South America*.

154. MURINE.   *Miller's plates*, xxxii. fig. ii. Lemur murinus, *Gm. Lin.* 44.

M. with head and body of an elegant light grey: infide of the ears white: orbits rufous: tail far exceeds the body in

* Pallas.

length:

length; bushy at the end, and of a bright rust color: nails flat and rounded: fize about twice that of a mouse.

Inhabits *Madagascar*, very nearly allied to the next: may only differ in sex.

Place.

Brown's Illuftr. of Zoology, 108. tab. xliv.

155. Little.

M. with a rounded head, sharp nose, long whiskers; two canine teeth in each jaw; four cutting teeth in the upper jaw; six in the lower: seven grinders on each side; the nearest sharp, the more distant lobated: the ears large, roundish, naked, and membranaceous: eyes very large and full.

The toes long, of unequal lengths; the ends round: nails round, and very short; that of the first toe strait, sharp, and long: tail hairy, of the length of the body, and is prehensile.

Color of the upper part cinereous; of the lower white; space round the eyes dark.

Rather less than the black rat.

Described from the living animal, in possession of Marmaduke Tunstal, *Esq.*

This seems to be the same animal, which M. *de Buffon* calls *Le Rat de Madagascar* *. It is supposed to live in the palm-trees, and feed on fruits. It eats, holding its food in its fore feet, like squirrels; is lively, and has a weak cry; when it sleeps, rolls itself up.

* *Supplem.* iii. 149. tab. xx.

Vol. I.   H h   Vespertilio

156. FLYING.

Vespertilio admirabilis. *Bontius Java*, 68.
Felis volans Ternatana. *Seb. Muf.* i. *tab.*
　　lviii.
Lemur volans. L. caudatus, membrana

ambiente volitans. *Lin. syst.* 45. *Schre-*
ber, 175. tab. xliii. LEV. MUS.
Galeopithecus *Act. Acad. Petrop.* 1780.
　p. 208. tab. vii.

M. with a long head: small mouth: small ears, round and membranous. No fore teeth in the upper jaw: six in the lower; short, broad, and elegantly pectinated, and distant from each other. From the neck to the hands, thence to the hind feet, extends a broad skin, like that of a flying squirrel; the same is also continued from the hind feet to the tip of the tail, which is included in it: the body and outside of this skin is covered with soft hairs, hoary or black, and ash-color; in adults the back is hoary, crossed transversely with black lines. The inner side of the extended skin appears membranous, with little veins and fibres dispersed thro' it: the legs are cloathed with a soft yellow down: five toes on each foot: the claws thin, broad, very sharp, and crooked, by which it strongly adheres to whatsoever it fastens on: the whole length of this species is near three feet; the breadth of the same: the tail slender; a span long.

Inhabits the country about *Guzarat*, the *Molucca* isles, and the *Philippines*: feeds on the fruits. Inhabits trees entirely. In descending from the top to a lower part it spreads its membranes, and balances itself to the place it aims at in a gentle manner; but in ascending uses a leaping pace. It has two young, which adhere to its breasts by its mouth and claws.

It is called by the *Indians, Caguang, Colugo,* and *Gigua.*

DIV.

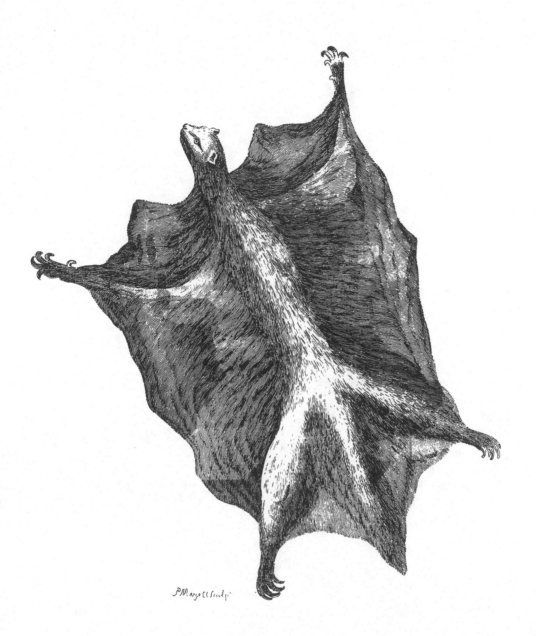

Flying . Maucauco __. N.º 156

DIV. II. Sect. II. Digitated Quadrupeds.

> With large canine teeth, feparated from the cutting teeth.
> Six, or more cutting teeth, in each jaw.
> Rapacious: carnivorous.

Six cutting teeth, and two canine, in each jaw.   XVII. DOG.
Five toes before; four behind *
Long vifage.

D. with its tail bending towards the left: a character com-   157. FAITHFUL.
mon to the whole fpecies; firft obferved by *Linnæus*.
Several beautiful varieties in the LEVERIAN MUSEUM.
The predominant paffion of the whole race towards an attachment to mankind, prevents thefe animals from feparating themfelves from us till deferted, or by fome accident left in places
where there was no poffibility of re-union: it feems beyond the
power of ill ufage to fubdue the faithful and conftant qualities inherent in them. Found in great numbers wild, or rather without
mafters, in *Congo*, *Lower Æthiopia*, and towards the *Cape* of *Good
Hope* †: are red-haired: have flender bodies, and turned-up tails,

---

* Invariable in the wild fpecies, fuch as wolf, &c.; in the common dogs, oft-
times five toes on each foot.

† *Churchill's coll. voy.* v. 486. *Kolben's hift. Cape*, ii. 106, 107.

like grehounds; others refemble hounds: they are of various colors, have erect ears, and are of the fize of a large fox-hound. Deftroy cattle, and hunt down antelopes as our dogs do the ftag *, and are very deftructive to the animals of chace: they run very fwiftly; have no certain refidence; are very feldom killed; being fo crafty as to fhun all traps: and of fo fagacious nofes as to avoid every thing that has been touched by man. Their whelps are fometimes taken; but grow fo exceffively fierce when they grow old, that they never can be domefticated.

They go in great packs: attack lions, tigers, and elephants, but are often killed by them: the fight of thefe dogs pleafing to travellers, who fuppofe they have conquered the wild beafts, and fecured their journey, by driving them away. Attack the fheep of the *Hottentots*, and commit great ravages among them.

Multitudes wild in *South America*: derived from the *European* race. Breed in holes, like rabbet-holes†: when found young, inftantly attach themfelves ‡ to mankind; nor will they ever join themfelves to the wild dogs; or defert their mafters: thefe have not forgot to bark ||, as *Linnæus* fays: look like a grehound §: have erect ears: are very vigilant: excellent in the chace.

The dog unknown in *America* before it was introduced there

---

* *Maffon*, in *Ph. Tranf.* lxvi. 278.

† Narrative of the diftreffes of *Ifaac Morris*, &c. belonging to the *Wager* florefhip, belonging to *Commodore Anfon's* fquadron, *p.* 27.

‡ The fame, *p.* 28.

|| The fame, *p.* 37.

§ As appears from a drawing communicated to me by Mr. *Greenwood*, painter, who took it from one that followed an *Indian* to *Surinam* from the inland part of the country.

by

by the *Europeans*: the *Alco* of the *Peruvians*, a little animal, which they were fo fond of, and kept as a lap-dog, is too flightly mentioned by *A-Cofta* for us to determine what it was: and the figure given by *Hernandez** too rude to form any judgment of: the other animal defcribed by *Hernandez* is a large fpecies, he calls it *Xoloitzicuintli*, the fame name that is given by the firft to the *Mexican* wolf†, as it is certain that the dog of *N. America*, or rather its fubftitute, on its firft difcovery by the *Englifh*, was derived from the ‡ wolf, tamed and domefticated; fo it is reafonable to imagine that of *S. America* had the fame origin. Thefe fubftitutes cannot bark, but betray their favage defcent by a fort of howl: want the fagacity of a true dog; ferve only to drive the deer into corners: the wolfifh breed to this day detefted ‖ by *European* dogs, who worry them on all occafions, retaining that diflike which it is well known all dogs have to the wolf. Thefe reclamed breed are commonly white: have fharp nofes, and upright ears.

The dog fubject to more variety than any other animal; each will mix with the other, and produce varieties ftill more unlike the original ftock. That of the old world is with great reafon fuppofed to be the *Schakal*, to which article the reader is referred. From the tamed offspring, again cafually croffed with the *Wolf*, the *Fox*, and even the *Hyæna*, has arifen the numberlefs forms and fizes of the canine race §. M. *de Buffon*, who with great ingenuity has given a genealogical table of all the known dogs, makes the *Chien de Berger*, the fhepherd's dog, or what is fometimes

---

* *Hernandez*, 466.   † *Hernandez*, 479.   ‡ *Smith's hift. Virginia*, 27. ‖ *Catefby Carolina*, ii. *App.* xxvi.   § *Pallas obf. fur la formation des Montagnes*, &c. 15.

called

called *Le Chien loup*, or the wolf dog, the origin of all, becaufe it is naturally the moft fenfible; becomes, without difcipline, almoft inftantly the guardian of the flocks; keeps them within bounds, reduces the ftragglers to their proper limits, and defends them from the attacks of the wolves. We have this variety in *England*; but it is fmall and weak. Thofe of *France* and the *Alps* are very large and ftrong; fharp-nofed, erect, and fharp-eared; very hairy, efpecially about the neck, and have their tails turned up or curled; and, by accident, their faces often fhew the marks of their combats with the wolf.

I fhall follow M. *de Buffon*, in the catalogue of dogs; but add fome few remarks, with the fynonyms of a few other writers, to each variety.

I. SHEPHERD's Dog, Le Chien de Berger. *De Buffon* *, v. 201. *tab.* xxviii. Canis domefticus. *Raii fyn. quad.* Lin. *fyft.* 57.

Its

* The *Englifh* reader will find all the varieties well defcribed and engraven in vol. iv. of Mr. *Smellie*'s tranflation of this author.

Notwithftanding M. *de Buffon* denies the junction of the wolf and bitch, yet there has been an inftance to the contrary. Mr. *Brook*, animal-merchant, in *Holborn*, turned a wolf to a *Pomeranian* bitch in heat: the congrefs was immediate, and as ufual between dog and bitch: fhe produced ten puppies. I have feen one of them, at *Gordon Caftle*, that had very much the refemblance of a wolf, and much of its nature; being flipped at a weak deer, it inftantly caught at the animal's throat and killed it. I could not learn whether this mongrel continued its fpecies: but another of the fame kind did; and ftocked the neighborhood of *Fochabers*, in the county of *Murray* (where it was kept) with a multitude of curs of a moft wolfifh afpect.

There

Its varieties, or neareſt allies, are,

α. POMERANIAN Dog, Le Chien Loup. *De Buffon, tab.* xxix.

β. SIBERIAN Dog, Le Chien de Siberie, *tab.* xxx. which is a variety of the former, and very common in *Ruſſia.* The other varieties, in the inland parts of the Empire and *Siberia,* are chiefly from the ſhepherd's dog: and there is a high-limbed taper-bodied kind, the common dog of the *Calmuc* and independent *Tartars,* excellent for the chace, and all uſes.

II. Hound, or dog with long, ſmooth, and pendulous ears. Le Chien courant, p. 205, *tab.* xxxii. Canis venaticus ſagax. *Raii ſyn. quad.* 177. Canis ſagax. *Lin. ſyſt.* 57. This is the ſame with the blood-hound. *Br. Zool.* i. 51. and is the head of the other kinds with ſmooth and hanging ears.

There was lately living a mongrel offspring of this kind. It greatly reſembled its wolf parent. It was firſt the property of Sir *Wolſtan Dixey:* afterwards of Sir *Willughby Aſton.* During day it was very tame; but at night ſometimes relapſed into ferocity. It never barked; but rather howled: when it came into fields where ſheep were, it would feign lameneſs, but if no one was preſent, would inſtantly attack them. It had been ſeen in copulation with a bitch, which afterwards pupped: the breed was imagined to reſemble in many reſpects the ſuppoſed ſire. It died between the age of five and ſix.

The bitch will alſo breed with the fox. The woodman of the manor of *Mongewell,* in *Oxfordſhire,* has a bitch, which conſtantly follows him, the offspring of a tame dog fox by a ſhepherd's cur: and ſhe again has had puppies by a dog. Since there are ſuch authentic proofs of the further continuance of the breed, we may ſurely add the wolf and fox to other ſuppoſed ſtocks of theſe faithful domeſtics.

α HARRIER.

D O G.

α HARRIER. Le Braque, *tab.* xxxiii.

β DALMATIAN*. Le Braque de *Bengal, tab.* xxxiv. a beautiful ſpotted kind, vulgarly called the *Daniſh* dog.

γ TURNSPIT. Le Baſſet a jambes torſes—a jambes droites, *tab.* xxxv.

δ WATER-dog, great and ſmall. Le grand and le petit Barbet, *tab.* xxxvii. xxxviii. Canis aviarius aquaticus. *Raii ſyn. quad.* 177. *Lin. ſyſt.* 57.

From N° II. branches out another race of dogs, with pendent ears, covered with long hairs, and leſs in ſize, which form

III. SPANIEL. Canis aviarius, five Hiſpanicus campeſtris. *Raii ſyn. quad.* 177. Canis avicularius? *Lin. ſyſt.* 57. Theſe vary in ſize, from the ſetting-dog to the ſpringing ſpaniels, and ſome of the little lap-dogs, ſuch as

α. KING CHARLES's †. Le Gredin, *tab.* xxxix. *fig.* 1.

---

* I have been informed, that *Dalmatia* is the country of this elegant dog. As for thoſe of *India*, they are generally ſmall and very ugly; or, if the *European* dogs are brought there, they immediately degenerate.

† CHARLES II. never went out, except attended by numbers of this kind.

β PYRAME.

β PYRAME. Le Pyrame, *tab.* xxxix. *fig.* 2. There is no *English* name for this kind: they are black, marked on the legs with red: and above each eye is a ſpot of the ſame color.

γ. SHOCK. Le Chien de *Malte* ou Bichon, *tab.* xl. *fig.* & Le Chien Lion, *fig.* 2. Catulus *melitæus* canis *getulus*, ſeu Iſlandicus. *Raii ſyn. quad.* 177. *Lin. ſyſt.* 57.

IV. Dogs with ſhort pendent ears: long legs and bodies: of which kind is the

α. IRISH GRE-HOUND. A variety once very frequent in *Ireland*, and uſed in the chace of the wolf: now very ſcarce: a dog of great ſize and ſtrength. Le Matin*. *De Buffon, tab.* xxv. Canis graius *Hibernicus*. *Raii ſyn. quad.* 176.

β. COMMON GRE-HOUND. Le Levrier. *De Buffon*, xxvii. *Schreber*, lxxxvii. Canis venaticus graius. *Raii ſyn. quad.* 176. Canis graius. *Lin. ſyſt.* 57. Its varieties are, 1. ITALIAN GRE-HOUND, ſmall, and ſmooth: 2. *Oriental*, tall, ſlender, with very pendulous ears, and very long hairs on the tail, hanging down a great length.

γ. DANISH DOG. Le grand Danois. *De Buffon*, xxvi. of a ſtronger make than a gre-hound: the largeſt of dogs: perhaps of this kind were the dogs of *Epirus*, mentioned by *Ariſtotle, lib.* iii. c. 21; or thoſe of *Albania*, the modern *Schirwan*, or *Eaſt Georgia*, ſo beautifully deſcribed by *Pliny, Lib.* viii. c. 40.

---

* Not the maſtiff, as commonly tranſlated.

Indian

Indiam petenti *Alexandro* magno, rex *Albaniæ* dono dederat inufitatæ magnitudinis unum; [*fcil. Canem*] cujus fpecie delectatus, juffit urfos, mox apros, et deinde damas emitti contemptu immobili jacente. Eaque fegnitie tanti corporis offenfus Imperator generofi fpiritus, eum interimi juffit. Nuntiavit hoc Fama Regi. Itaque alterum mittens addidit mandata, ne in parvis experiri vellet, fed in leone, elephantove. Duos fibi fuffe: hoc interèmpto, præterea nullum fore. Nec diftulit *Alexander,* leonemque fractum protinus vidit. Poftea elephantum juffit induci, haud alio magis fpectaculo lætatus. Horrentibus quippe per totum corpus villis, ingenti primum latratu intonuit. Mox ingruit affultans, contraque belluam exurgens hinc & illinc, artifici dimicatione, qua maxime opus effet, infeftans atque evitans, donec affidua rotarum vertigine afflixit, ad caíum ejus tellure concuffa.

Perhaps to this head may be referred the vaft dogs of *Thibet,* faid by *Marco Polo* to be as big as affes, and ufed in that country to take wild beafts, and efpecially the wild oxen called *Beyamini* *.

♂. MASTIFF. Very ftrong and thick made: the head large: the lips great, and hanging down on each fide: a fine and noble countenance: grows to a great fize: a *Britifh* kind. For a further account of this and other *Britifh* dogs, *vide Br. Zool.* i. 49. Le Dogue de forte race. *De Buffon, tab.* xlv. Maftivus. *Raii fyn. quad.* 176. Canis moloffus. *Lin. fyft.* 57.

V. Dogs with fhort pendent ears: fhort compact bodies: fhort nofes: and generally fhort legs.

♀. BULL-DOG: with a fhort nofe, and under jaw longer than the upper: a cruel and very fierce kind, often biting before it barks: peculiar to *England:* the breed fcarcer than it has been,

* *Purchas,* iii. 90.

4.

since the barbarous custom of bull-baiting has declined. Le Dogue. *De Buffon, tab.* xliii.

β. Pug Dog. A small species: an innocent resemblance of the last. Le Doguin. *De Buffon, tab.* xliv.

γ. Bastard Pug. Le Roquet. *De Buffon,* xli. *fig.* 2.

δ. Naked. Le chien Turc. *De Buffon,* xlii. a degenerate species, with naked bodies; having lost its hair by the heat of climate.

Dogs (brought originally from *New Guinea*)*, are found in the *Society Islands, New Zeland,* and the *Low Islands*: there are also a few in *New Holland.* Of these are two varieties.

Dogs of the S. Sea Islands.

1. Resembling the sharp-nosed pricked-ear shepherd's cur. Those of *New Zeland* are of the largest sort. In the *Society Islands* they are the common food, and are fattened with vegetables, which the natives cram down their throats, as we serve turkies, when they will voluntarily eat no more. They are killed by strangling, and the extravasated blood is preserved in *Coco*-nut shells, and baked for the table. They grow very fat, and are allowed, even by *Europeans* who have got over their pre-judices, to be very sweet and palatable.

Eaten there.

But the taste for the flesh of these animals was not confined to the islanders of the *Pacific Ocean.* The antients reckoned a young and fat dog excellent food, especially if it had been castrated † : *Hippocrates* placed it on a footing with mutton and pork ‡ : and in another place says, that the flesh of a grown dog is wholesome and strengthening; of puppies (if I take him right)

Flesh of dogs eaten by the antients.

---

* See this edition under title Hog.   † *Galen,* lib. iii. de Alim. facul . c. 11.

‡ De intern. affect. Sect. v.

I i 2

relaxing.

relaxing*. The *Romans* admired fucking puppies : they facrificed them alfo to their divinities, and thought them a fupper in which the Gods themfelves delighted †.

2. The *Barbet,* whofe hair being long and filky, is greatly valued by the *New Zelanders* for trimming their ornamental drefs. This variety is not eaten. The iflanders never ufe their dogs for any purpofes but what we mention ; and take fuch care of them as not to fuffer them even to wet their feet. They are exceffively ftupid, have a very bad nofe for fmelling, and feldom or never bark, only now and then howl. The *New Zelanders* feed their dogs entirely with fifh.

The *Marquefas, Friendly Iflands, New Hebrides, New Caledonia,* and *Eafter Ifle,* have not yet received thofe animals.

‡ The moft faithful of animals: is the companion of mankind: fawns at the approach of its mafter: will not fuffer any one to ftrike him: runs before him in a journey ; often paffing backward and forward over the fame ground : on coming to crofs-ways, ftops and looks back : very docile : will find out what is dropt: watchful by night: announces the coming of ftrangers: guards any goods committed to its charge: drives cattle home from the field: keeps herds and flocks within bounds : protects them from wild beafts: points out to the fportfman the game, by virtue of its acute fenfe of fmelling: brings the birds that are fhot to its mafter: will turn a fpit: at *Bruffels* and in *Holland* draws little carts to the herb-market: in *Siberia* draws a fledge with its mafter in it, or loaden with provifions: fits up

* De Diæt. et facult. lib. ii.
† *Plin. hift. lib.* xxix. c. iv.
‡ This part is almoft entirely tranflated from *Linnæus.*

and

and begs*: when it has committed a theft, flinks away with its tail between its legs: eats envioufly, with oblique eyes: is mafter among its fellows: enemy to beggars: attacks ftrangers without provocation†: fond of licking wounds: cures the gout and cancers:

* The *French* Academicians record a marvellous tale of a dog that could *fpeak*, and call for tea, coffee, and chocolate.

† This part of the nature of dogs is fo elegantly expreffed by *Theocritus*, that the reader will not be difpleafed with the reference, and the tranflation by the Rev. Mr. *Fawkes*, giving an account of the inftinct of the old herdfman's dogs at the approach of *Hercules*.

Τὰς δε κυνες προσιονίας αποπρωσθεν αιψ' ενοησαν,
Αμφοτερον οδμη τε χροος, δαπω τε ποδοιιν.
Θεσπεσιον δ' υλαοντες επεδραμον αλλοθεν αλλω·
Αμφιτρυωνιαδη Ηρακλει· τον δε γεροντα
Αχρειον κλαζοντε, περισσαινον θ' ετερωθεν.
Τὰς μεν ο ζε λαεσσιν, απο χθονος οσσον αειρων,
Φευζεμεν αψ οπισω δειδισσετο· τρηχυ δε φωνη
Ηπειλει μαλα πασιν, ερητυσασκε δ' υλαζμα·
Χαιρων εν φρεσιν ησιν, οθ' αυνεκεν αυλιω ερυντο,
Αυτα γ' α παρεολω· επω δ' ο ζε τοιον εειπεν·
Ω ποποι, οιον τατο θεοι ποιησαν ανακιες
Θηριον ανθρωποισι μελεμμεναι· ως επιμηθες.
Ει οι και φρενες ωδε νοημονες ενδοθεν ησαν,
Η δει δ' ω ζε χρη χαλεπαινεμεν, ωζε και ακι,
Ουκ αν τοι θηρων τις εδηρισεν περι τιμης·
Νυν δε λιην ζακολον τε και αρρενες γενεθ' αυλως.
Η ρα· και εσσυμενως πολι ταυλιον ιζον ιονες.

Idyl. XXV. *v.* 68.

The watchful dogs, as near the ftalls they went,
Perceiv'd their coming by their tread and fcent,
With open mouths from every part they run,
And bay'd inceffant great *Amphitryon's* fon;

But

cers: howls at certain notes in mufic, and often urines on hearing them: bites at a ftone flung at it: is fick at the approach of bad weather: gives itfelf a vomit by eating grafs: is afflicted with tape-worms: fpreads its madnefs: grows blind with age: *fæpe gonorrhæa infectus:* driven as unclean from the houfes of the *Mahometans*; yet the fame people eftablifh hofpitals for them, and allow them a daily dole of food: eats flefh, carrion, farinaceous vegetables, not greens: fond of rolling in carrion: dungs on a ftone; its dung the greateft of *feptics:* drinks by lapping: makes water fide-ways, with its leg held up; very apt to repeat it where another dog has done the fame: *odorat anum alterius: menftruans catulit cum variis; mordet illa illos; cohæret copula junctus.* Goes 63 days with young; brings from four to ten; the males like the dog, females like the bitch: its fcent exquifite: goes obliquely: foams when hot, and hangs out its tongue: fcarce fweats: about to lie down, often goes round the fpot: its fleep attended with a quick fenfe of hearing: dreams.

But round the fwain they wagg'd their tails and play'd,
And gently whining, fecret joy betray'd;
Loofe in the ground the ftones that ready lay
Eager he fnatch'd, and drove the dogs away;
With his rough voice he terrify'd them all,
Though pleas'd to find them guardians of his ftall.
' Ye Gods! (the good old herdfman thus began)
' What ufeful animals are dogs to man!
' Had Heav'n but fent intelligence to know
' On whom to rage, the friendly or the foe,
' No creature then could challenge honour more;
' But now too furious and too fierce they roar.'
He fpoke, the growling maftives ceas'd to bay,
And ftole obfequious to their ftalls away.

Stockdale's

Stockdale's Bot. Bay, 274. *White's,* 280.

158. NEW HOLLAND.

D. with short erect sharp-pointed ears: a fox like head; color of the upper part of the body pale brown; grows lighter towards the belly; hind part of the fore legs, and fore part of the hind legs, white: feet of both of the same color: tail very bushy: length about two feet and a half: of the tail not a third of that of the body: height about two feet.

Inhabits *New Holland,* and seems the unreclamed dog of the country. Two have been brought alive to *England;* are excessively fierce, and do not shew any marks of being brought to a state of domesticity. It laps like other dogs; but neither barks or growls, when provoked; but erects its hair like bristles, and seems quite furious. Is eager after its prey; and is fond of rabbets and fowls, but will not touch dressed meat: is very agile: It once seized on a fine *French Dog* by the loins, and would have soon destroyed it had not help been at hand. It leaped with great ease on the back of an ass, and would have worried it to death had not the ass been relieved, for it could not disengage itself from the assailant. It was known to run down deer and sheep.

PLACE.

MANNERS.

Lupus.

159. Wolf.

Lupus. *Gesner quad.* 634.   *Raii syn.*   *syst.* 58.
   quad. 173.
Wolf. *Klein quad.* 69. *Kram. Aust.* 313.
Canis ex griseo flavescens. *Brisson quad.*
   170.
Canis Lupus. C. cauda incurvata. *Lin.*

Warg, Ulf. *Faun. suec.* No. 6.
Le Loup. *De Buffon,* vii. 39. *tab.* i.
Wolf. *Br. Zool.* i. 62. *tab.* 5. *Schreber,*
   lxxxviii. Lev. Mus.

D. with a long head: pointed nose: ears erect and sharp: tail bushy, bending down; the tip black: long leg'd: hair pretty long: teeth large: head and neck cinereous: body generally pale-brown, tinged with yellow; sometimes found white*: taller than a large grehound. In *Canada* sometimes black: and called by *Linnæus, Canis-Lycaon.*

Place.

Inhabits the continents of *Europe, Asia,* and *America; Kamtschatka,* and even as high as the *Arctic* circle. Is unknown in *Africa,* notwithstanding M. *Adanson*† speaks of it familiarly. The *French* and other naturalists mistake the *Hyæna* for this animal. Has been long extirpated in *Great Britain*‡. The last wolf which was known in this island, was killed in *Scotland* in 1680, by the famous Sir *Ewen Cameron,* according to the tradition of the country. I have travelled into almost every corner of that country; but could not learn that there remained even the memory of those animals among the oldest people. In *Ireland* they continued longer; for one was killed in that island in 1710, when the last pre-

* Such are found near the *Jenesea,* and sold to the *Russians* on the spot for twenty shillings a skin. *Muller Russ. Samlung.* iii. 527, 529.
† P. 209.
‡ M. *de Buffon* must have been greatly misinformed on this point. *Les Anglois* pretendent *en avoir purgè leur isle, cependant on m'a assurè qu'il y en avoit en* Ecosse, vii. 50.

5   sentment

fentment for killing of wolves was made in the county of *Cork**.
In 1281, I find that they infefted feveral of the *Englifh* counties †;
but after that period, our records make no mention of them. The
vaft forefts on the continent of *Europe* will always preferve them.

The wolves of *N. America* the fmalleft; when reclamed, are
the dogs of the natives.

Are cruel, but cowardly animals: fly from man, except preffed
by hunger, when they prowl by night in vaft droves thro' villages, and deftroy any perfons they meet: fuch as once get the
tafte of human blood, give it the preference: fuch were the
wolves of the *Gevaudan,* of which fo many ftrange tales were
told: the *French* peafants call this *Loup-garou,* and fuppofe it to
be poffeffed with fome evil fpirit: fuch was the *Were Wulf* of the
old *Saxon* ‡. The wolf preys on all kinds of animals; but in
cafe of neceffity will feed on carrion: in hard weather affemble in
vaft troops, and join in dreadful howlings: horfes generally defend themfelves againft their attacks; but all weaker animals fall
a prey to them: throughout *France* the peafants are obliged
nightly to houfe their flocks. Wolves are moft fufpicious animals;
fally forth with great caution: have a fine fcent; hunt by nofe:
are capable of bearing long abftinence: to allay their hunger will
fill their bellies with mud: a mutual enmity fubfifts between dogs
and them: are in heat in winter; followed by feveral males,
which occafions great combats: goes with young ten weeks:
near her time, prepares a foft bed of mofs, in fome retired place:
brings from five to nine at a birth: the young born blind: teeth

* *Smith's hift. Cork*, ii. 226. † *Rymer's Fœd.* ii. 168.
‡ *Verftegan's Antiq.* 236.

K k of

of the wolf large and sharp : its bite terrible, as its strength is great : the hunters therefore clothe their dogs, and guard their necks with spiked collars. Wolves are proscribed animals : destroyed by pit-falls, traps, or poison : a peasant in *France*, who kills a wolf, carries its head thro' the villages, and collects some small reward from the inhabitants : the *Kirghis-Khaissacks* take the wolves by the help of a large sort of hawk called *Berkut*, which is trained for the diversion, and will fasten on them and tear out their eyes *.

160. MEXICAN.

MEXICAN WOLF. Xoloitzcuintli. *Hernandez Mex.* 479.
Cuetlatchtli, seu lupus Indicus. *Hernandez An. Nov. Hisp.* 7.
Canis cinereus, maculis fulvis variegatus, tæniis subnigris a dorso ad latera deorsum hinc inde deductis. *Brisson quad.* 172.
Canis Mexicanus. C. cauda deflexa lævi, corpore cinereo, fasciis fuscis, maculisque fulvis variegato. *Lin. syst.* 60.
Le Loup de Mexique. *De Buffon,* xv. 149.

D. with a very large head : great jaws : vast teeth : on the upper lips very strong bristles, reflected backwards, not unlike the softer spines of a porcupine ; and of a grey and white color : large, erect, cinereous ears ; the space between marked with broad tawny spots : the head ash-colored, striped transversely with bending dusky lines : neck fat and thick, covered with a loose skin, marked. with a long tawny stroke : on the breast is another of the same kind : body ash-colored, spotted with black ; and the sides striped, from the back downwards, with the same color : belly cinereous : tail long, of the color of the

* *Rischkoff Topog. Orenb.* i. 282.

belly,

belly, tinged in the middle with tawny: legs and feet ſtriped with black and aſh-color: ſometimes this variety (for *Hernandez*, who has deſcribed the animals of *Mexico*, thinks it no other) is found white.

Inhabits the hot parts of *Mexico*, or *New Spain*: agrees with the *European* wolf in its manners: attacks cattle, and ſometimes men. No wolves found farther ſouth, on the new continent.

Vulpes. *Geſner quad.* 966. *Raii ſyn. quad.* 177.
Fuchs. *Klein quad.* 73. *Meyer's An.* i. *tab.* 36.
Canis vulpes. C. cauda recta apice albo. *Lin. ſyſt.* 59. *Haſſelquiſt, itin.* 191.

Raef. *Faun. ſuec.* No. 7.
Canis fulvus, pilis cinereis intermixtis. *Briſſon quad.* 173.
Le Renard. *De Buffon*, vii. 75. *tab.* vi.
Fox. *Br. Zool.* i. 58. Lev. Mus.

161. Fox.

D. with a ſharp noſe: lively hazel eyes: ſharp erect ears: body tawny red, mixed with aſh-color: fore part of the legs black: tail long, ſtrait, buſhy, tipt with white: ſubject to much variety in color.

α. Fox: with the tip of the tail black. Canis alopex, vulpes campeſtris. *Lin. ſyſt.* 59.

β. Cross Fox: with a black mark, paſſing tranſverſely from ſhoulder to ſhoulder; and another along the back, to the tail. Vulpes crucigera. *Geſner quad.* 90. *Jonſton quad.* i. 93. *Schæffer Lapl.* 135. *Hiſt. Kamtſchatka.* 95. *Klein quad.* 71.
Le Renard croiſé. *Briſſon quad.* 173. *De Buffon*, xiii. 276.
Korſraef. *Faun. ſuec.* p. 4.

K k 2

Inhabits

Inhabits the coldeſt parts of *Europe*, *Aſia*, and *North America:* a valuable fur; thicker and ſofter than the common ſort: great numbers of the ſkins imported from *Canada*. Not a variety of the *Iſatis* or *Arctic* fox.

γ. BLACK FOX. The moſt cunning of any: and its ſkin the moſt valuable; a lining of it eſteemed in *Ruſſia* preferable to that of the fineſt ſables: a ſingle ſkin will ſell for 400 rubles. Inhabits the northern parts of *Aſia*, and *N. America:* the laſt of inferior goodneſs.

δ. BRANT FOX. That deſcribed by *Geſner** and *Linnæus†* is of a fiery redneſs; and called by the firſt *Brand-fuſchs*, by the laſt *Brandraef:* one that was the property of Mr. *Brook*, was ſcarcely half the ſize of the common fox: the noſe black, and much ſharper: ſpace round the ears ferruginous: forehead, back, ſhoulders, ſides, and thighs, black, mixed with red, aſh-color, and black; the aſh-color predominated, which gave it a hoary look: the belly yellowiſh: tail black above, red beneath: cinereous on its ſide. This Mr. *Brook* received from *Penſylvania*, under the name of Brant fox.

2. KARAGAN.　　Allied to this is the *Karagan*, a ſmall ſpecies very common in all parts of the *Kirghiſian* deſerts and *Great Tartary*.

Head yellowiſh above, reddiſh above the eyes: behind the whiſkers is a black ſpot: ears black without; white within: exterior edge and baſe red; and near the baſe of that edge is a white

---

* *Geſner quad.* 967, who likewiſe ſays, it is leſs than the common kind.
† *Faun. ſuec.* No. 7.

8

ſpot:

fpot: the color of the back and fides like a wolf; and the hair
coarfe in the fame degree: between the fhoulders is a dark fpot,
from which, along the back to the tail, extends a reddifh or yellowifh track: a deep grey or blackifh fpace, mixed with white,
covers the throat, and is continued over the breaft and part of
the belly; the reft of which is whitifh.

A fmall kind, defcribed to me by Doctor *Pallas* from the
fkins.

ζ. Corsak Fox. Canis corfac. C. cauda fulva bafi apiceque nigra. *Lin. fyft.* iii.
223. *Schreber*, xci. B.

D. with upright ears: foft downy hair: tail bufhy, the length
of the body: throat white: irides yellowifh green: color
in fummer pale tawny; in winter grey: hair coarfer than that of
the common fox: bafe and tip of the tail black; the reft cinereous: is a fmall fpecies.

Inhabits the deferts beyond the *Yaik*; and from the *Don* to
the *Amur:* lives in holes, and burrows deep: howls and barks:
never found in woody places: caught by the *Kirghis-Khaiffacks*,
with falcons and gre-hounds: 40 or 50,000 are taken annually,
and fold to the *Ruffians*, at the rate of 40 *Kopeiks*, or 20 pence
each. The former ufe their fkins inftead of money. Great numbers are fent into *Turky* *.

M. *de Buffon* confounds this with the *Ifatis*, or *Arctic* Fox †.

* *Ritchkoff Topogr. Orenb.* i. 296.
† Suppl. iii. p. 113. tab. xvii.

COMMON

COMMON Fox inhabits all *Europe*, the cold and temperate parts
of *Asia*, *Barbary*, but not the hotter parts of *Africa*; abounds in
N. America; and also found in *S. America* *: in all countries they
have the same cunning disposition; the same eargerness after
prey; and commit the same ravages among game, birds, poultry, and the lesser quadrupeds: are very fond of honey; attack
the wild bees and nests of wasps, for the sake of the maggots: will
eat any sort of insects: devour fruit; and are very destructive in
vineyards: bury what they cannot eat: fond of basking in the sun.

Lodge under ground; generally making use of a badger's
hole, which they enlarge, adding several chambers, and never
neglecting to form another hole to the surface to escape at, in
cases of extremity: prey by night: females in heat in winter;
bring five or six at a time; if the young are disturbed, will remove them one by one to a more secure place: their voice a
yelp, not a bark: their bite, like that of the wolf, is very hard
and dangerous: their scent excessively strong; the chace on that
account more keen, more animating: when chased, first attempt
to recover their hole, but finding that stopt, generally fly the
country.

These animals are extremely common in the Holy Land †.
From the earliest to the present time, they were particularly
noxious to the vineyards; " Take us the *foxes*, the little foxes
" that spoil the vines; for our vines have tender grapes ‡."
Whether they were the species of which *Sampson* made use, to

* *Garcilasso de la Vega* says, that the foxes of *Peru* are much less than those of
Spain, and are called *Atoc*. P. 331.

† *Hasselquist*, Original 191. Transl. 184.

‡ Song of SOLOMON, ii. 15.

destroy

Arctic Fox —— N.º 162.

deſtroy the corn of the *Philiſtines*, is undecided. Since *Schakals* are found to this day in great abundance about *Gaza* *, it is much more probable, from their gregarious nature, that he could catch three hundred of them, than of the ſolitary quadruped the fox.

Vulpes alba. *Jonſton quad.* 93.
Fox. *Marten's Spitzberg.* 100. *Egede Greenl.* 62. *Crantz Greenl.* i. 72.
Aſhen-colored Fox. *Schæffer Lapland,* 135.
Canis Lagopus. C. caudata recta, apice concolore. *Lin. ſyſt.* 59.

Fial racka. *Faun ſuec.* No. 8.
Canis hieme alba, æſtate ex cinereo cærulescens. *Briſſon quad.* 174. *Schreber*, xciii.
Iſatis. *Nov. Com. Petrop.* v. 358. *De Buffon*, xiii. 271. *Aſp. Muſ.* Lev. Mus.

162. ARCTIC.

D. with a ſharp noſe: ſhort rounded ears, almoſt hid in the fur: long and ſoft hair, ſomewhat woolly, and of a white color; ſometimes pale cinereous: ſhort legs: toes covered on all parts, like thoſe of a hare, with fur: tail ſhorter than that of a common fox, and more buſhy: hair much longer in winter than ſummer, as uſual with animals of cold climates.

Inhabits the countries bordering on the frozen ſea, as far as the land is deſtitute of woods, which is generally from 70 to 68 degrees latitude. The ſpecies extends to *Kamtſchatka*, and in *Bering's* and *Copper Iſlands*, but in none of the other iſlands between *Kamtſchatka* and the oppoſite parts of *America*, diſcovered in Captain *Bering's* expedition, 1741; is again found in *Greenland, Iceland, Spitzbergen, Nova Zembla*, and *Lapland*: burrows

PLACE.

* *Haſſelquiſt.*

under

under ground; forms holes many feet in length; ſtrews the bottom with moſs; in *Greenland* and *Spitzbergen*, lives in the cliffs of rocks, not being able to burrow, by reaſon of the froſt: two or three pair inhabit the ſame hole: are in heat about *Lady-Day*; during that time continue in the open air; afterwards take to their holes: go with young nine weeks: like dogs, continue united in copulation: bark like that animal; for which reaſons the *Ruſſians* call them *Peſzti* or dogs: have all the cunning of the common fox: prey on the young of geeſe, ducks, and other water-fowl before they can fly; on grouſe of the country, and hares; on the eggs of birds; and in *Greenland* (through neceſſity) on berries, ſhell-fiſh, or any thing the ſea flings up: but their principal food in the *North* of *Aſia*, and in *Lapland*, is the *Leming*: thoſe of the countries laſt mentioned are very migratory, purſuing the *Leming*, a very wandering animal: ſometimes theſe foxes will deſert the country for three or four years, probably in purſuit of their prey; for it is well known that the migrations of the *Leming* are very inconſtant, appearing in certain countries only once in ſeveral years: the people of *Jeneſea* ſuſpect they go to the banks of the *Oby*. Are taken in traps: oft-times the *Glutton* and Great owl deſtroy them, before the hunter can take them out: the ſkins of ſmall value. The great rendezvous of theſe animals is on the banks of the frozen ſea, and the rivers that flow into it, being found there in great troops. *Molina* found this ſpecies in *Chili* *.

* P. 253.

Arct.

Arɕt. Zool. i. p. 90.

With a duſky fur on every part; in ſize and habit reſembling the former.

A diſtinɕt ſpecies. Inhabits *Iceland* in great numbers. Communicated to me by *John Thomas Stanley*, Eſq.

Arɕt. Zool. i. 91.

Above of a ſooty brown: ears rounded, white within: a white bed extends from each to the lower part of the throat, which, with the whole underſide, and inſide of the haunches, is white: tail white below, brown above; in one ſpecimen the one half of the tail wholly white: beneath each eye a white ſpot: feet furred beneath. A very ſmall ſpecies.

Inhabits *Greenland.* Bought by Mr. *Stanley* at *Copenhagen.*

Coyotl ſeu vulpes Indica.   *Hernandez Anim. Mex.* 4.
Loup-renard. Wolf fox.   *Bougainville's voy. tranſl.* 58.

D. with ſhort pointed ears; their inſides lined with white hairs: irides hazel: head and body cinereous brown: hair more woolly than that of the common fox, reſembling much that of the *arɕtic*: legs daſhed with ruſt-color: tail duſky, tipped with white; ſhorter and more buſhy than that of the common fox, to which it is about one-third ſuperior in ſize. It has much the habit of the wolf, in ears, tail, and ſtrength of limbs. The *French* therefore call it *Loup-renard*, or Wolf-fox. It may be a wolf

degenerated by climate. The largeſt are thoſe of *Europe:* thoſe of
North America are ſtill ſmaller. The *Mexican* wolves, which I
apprehend to be this ſpecies, are again leſs ; and this, which inhabits the *Falkland* iſles, near the extremity of *South America,* is
dwindled to the ſize deſcribed.

It is the only land animal of thoſe diſtant iſles : lives near the
ſhores: kennels like a fox ; and forms regular paths from bay to
bay, probably for the conveniency of ſurprizing the water-fowl,
on which it lives. It is at times very meager, for want of prey :
is very tame; fetid, and barks like a dog.

The iſlands were probably ſtocked with thoſe animals by
means of iſlands of ice broken from the continent, and carried by
the currents.

This deſcription was taken from one brought to *England* when
we poſſeſſed thoſe *antarctic* ſpots. The following ſeems only a
variety of this ſpecies.

<table><tr><td>166. A. CULPEU.</td><td>Canis culpæus, *Molina Chili.* 274.</td></tr></table>

D. with a ſtrait tail, covered with ſhort hair, like the domeſtic
dog: color deep brown. In all reſpects of form reſembles the fox, but is larger: length to the tail two feet and a
half.

Inhabits the open countries of *Chili,* in which it forms its
boroughs. Its voice is feeble, but has ſome reſemblance of barking. If it ſees a man at a diſtance, will march ſtrait towards him;
ſtop at a diſtance, and regard him attentively. If the man makes
no movement, will remain long in the ſame ſituation, but without

MANNERS.

5

out doing him the left harm, and then retires the same way it came.
This *Molina* often had occasion to remark: for it never failed
doing the same thing.   This subjects it to the shot of the sports-
men: the *Chilians* call it *Culpeu* from *Culpem*, which signifies
folly.

This is certainly the same as the foregoing. Mr. *Byron* * found
them in great numbers on *Falkland* isles.   They constantly came
running up to the men, which was mistaken for a design to attack
them; which it does not appear these animals ever did.

<table>
<tr><td>Canis cinereo-argenteus, *Erxleb.* 567. *Schreber,* tab. xcii. 4.</td><td>167. SCHRE-BERIAN.</td></tr>
</table>

D. with its neck and sides tawny; ears tawny within, tipt
 • with black: crown and back mixed with grey, black, and
white : throat, breast, and belly, white: less than the common fox.
Inhabits *North America*.   Possibly the young of the preceding.

<table>
<tr><td>Grey fox. *Smith's voy. Virginia,* 27. *Josselyn's voy.* 81. *Rarities,* 21. *Law-son's Carolina,* 125. *Catesby Carolina,* ii. 78. Canis ex cinereo argenteus. *Brisson quad.* 174. *Schreber,* xci. xcii.</td><td>168. GREY.</td></tr>
</table>

D. with a sharp nose: sharp, long, upright ears: legs long:
 • color grey, except a little redness about the ears.
Inhabits *Carolina*, and the warmer parts of *N. America*: differs
from the arctic fox in form; and in the nature of its dwelling: agrees
with the common fox in the first, varies from it in the last : never
burrows; lives in hollow trees: gives no diversion to the sports-
man, for after a mile's chace takes to its retreat: has no strong

* Voyage in *Hawkesworth's* coll. i. 49, 50.

L l 2   smell:

smell: feeds on poultry, birds, &c.: eafily made tame: their fkins, when in feafon, made ufe of for muffs.

169. Silvery.　Le Renard argentè. *Charlevoix Nouv. France,* v. 196. *Du Pratz, Louifian.* ii. 64.

IN form refembling the common fox: abound in the wooded eminences in *Louifiana,* which are every where-pierced with their holes: their coat very beautiful: the fhort hairs of a deep brown; over them fpring long filvery hairs, which give the animal a very elegant appearance: they live in forefts abounding in game, and never attempt the poultry which run at large.

170. Bengal.　D. of a light brown color: face cinereous, with a black ftripe down the middle, and a white fpace round the eyes and middle of the jaws; with fulvous legs: tail tipt with black: a fpecies fcarcely half the fize of the *European* fox.

Inhabits *Bengal:* feeds chiefly on roots and berries. The *Englifh,* at a vaft expence, import into *India* hounds for the purpofe of the chace; which quickly degenerate.

171. Barbary.　Le Chacal. *De Buffon,* Supplem. vi. 112. *tab.* xvi.

D. with a long and flender nofe, fharp upright ears, long bufhy tail: color a very pale brown: fpace above and below the eyes black: from behind each ear is a black line, which foon divides into two, which extend to the lower part of the neck: the tail furrounded with three broad rings: fize of the common fox, but the limbs fhorter, and the nofe more flender.

I had

I had a drawing made from the skin of this animal, badly preserved, some years ago, in the *Aſhmolean Muſeum, Oxford*, which I sent to M. *de Buffon.* He cauſed it to be engraven; and informs us that Mr. *Bruce* told him it was common in *Barbary*, where it was called *Thaleb.* Mr. *Bruce* ſhould have given it a more diſtinguiſhing name; for *Thaleb* *, or *Taaleb*†, is no more than the *Arabic* name for the common fox, which is alſo frequent in that country.

Adil, Squilachi, *Græc. modern. Belon, obſ.* 163.
Lupus Aureus. *Kæmpfer Amœn. exot.* 413. *Raii ſyn. quad.* 174. *Klein quad.* 70.
Canis aureus. *Lin. ſyſt.* 59.
Canis flavus. *Briſſon. quad.* 171.
Le Chacal & L'Adive. *De Buffon,* xiii.

255. *Zimmerman,* 361. *Schreber,* xciv. 472. Lev. Mus.
Schakali. Hiſt. *Gueldenſtaedt,* in Nov. Com. *Petrop.* xx. 449. *tab.* xi.
Vaui, *ou* Benat el Vaui. *Niebuhr deſcr. Arab.* 146. Lev. Mus.

172. Schakal.

D. with yellowiſh brown irides; ears erect, formed like thoſe of a fox, but ſhorter and leſs pointed: hairy and white within; brown without, tinged with duſky: head ſhorter than that of a fox, and noſe blunter: lips black, and ſomewhat looſe: neck and body very much reſembling thoſe of that animal, but the body more compreſſed: the legs have the ſame reſemblance, but are longer: tail thickeſt in the middle, tapering to the point: five toes on the fore feet; the inner toe very ſhort, and placed high: four toes on the hind feet; all are covered with hair even to the claws.

The hairs much ſtiffer than thoſe of a fox, but ſcarcely ſo ſtiff

* *Shaw's travels,* 249.   † *Forſkal's obſ. p.* 111.

8

as

as thofe of a wolf; fhort about the nofe; on the back three inches long; on the belly fhorter. Thofe at the end of the tail four inches long.

COLOR.

Color of the upper part of the body a dirty tawny; on the back mixed with black: lower part of the body of a yellowifh white: tail tipt with black; the reft of the fame color with the back: the legs of an unmixed tawny brown; the fore legs marked (but not always) with a black fpot on the knees; but on no part are thofe vivid colors which could merit the title of *golden*, beftowed on it by *Kæmpfer*.

I avoid in general the mention of the internal ftructure of animals, from a confcioufnefs of my deficiency in that branch of fcience: but muft here remark from Profeffor GUELDENSTAEDT, the able defcriber of this long-loft animal, that the *cæcum* entirely agrees in form with that of a dog, and differs from that of the wolf and fox. I may add, that there is the fame agreement in the teeth with thofe of a dog; and the fame variation in them from thofe of the two other animals. I mention this, as it is an opinion with fome writers, that the dogs of the old world did derive their origin from one or other of them.

SIZE.

The length of the *Schakal*, from the nofe to the root of the tail, is little more than twenty-nine inches *Englifh*: the tail, to the ends of the hairs, ten three quarters, the tip reaching to the top of the hind legs: the height, from the fpace between the fhoulders to the ground, rather more than eighteen inches and a half; the hind parts a little higher.

PLACE.

Inhabits all the hot and temperate parts of *Afia, India, Perfia, Arabia, Great Tartary,* and about Mount *Caucafus, Syria,* and the *Holy-land.* In moft parts of *Africa,* from *Barbary* to the *Cape of Good Hope.*

They

They have so much the nature of dogs, as to give reasonable cause to imagine that they are (at least) the chief stock from which is sprung the various races of those domestic animals. When taken young, grow instantly tame; attach themselves to mankind; wag their tails; love to be stroked; distinguish their masters from others: will come on being called by the name given to them; will leap on the table, being encouraged to it: drink lapping: make water sideways, with their leg held up. Their dung hard: *odorat anum alterius, cohæret copula junctus.* When they see dogs, instead of flying, seek them, and play with them *: will eat bread eagerly; notwithstanding it is in a wild state carnivorous: has a great resemblance to some of the *Calmuc* dogs, which perhaps were but a few descents removed from the wild kinds. Our dogs are probably derived from those reclamed in the first ages of the world; altered by numberless accidents into the many varieties which now appear among us. The wild *Schakals* go in packs of forty, fifty, and even two hundred, and hunt like hounds in full cry from evening to morning †. They destroy flocks and poultry, but in a less degree than the wolf or fox: ravage the streets of villages, and gardens near towns, and will even destroy children ‡ if left unprotected. They will enter stables and outhouses, and devour skins, or any thing made of that material: are bold thieves; will familiarly enter a tent, and steal whatsoever they can find from the sleeping traveller. In default of living prey, they will feed on roots and fruits; and even on the most infected carrion: will greedily dis-

---

* *Nov. Com. Petrop.* xx. 459. *Pallas, Sp. Zool. fasc.* xi. 1.
† *Belon obs.* 163. ‡ *Dellon's voy.* 81.

inter the dead *, and devour the putrid carcafes; for which reafon, in many countries, the graves are made of a great depth.
They attend caravans, and follow armies, in hopes that death will
provide them a banquet.

Their voice naturally is a howl. Barking is latently inherent;
and in their ftate of nature feldom exerted: but its different mo-
difications are adventitious, and expreffive of the new paffions and
affections gained by a domeftic ftate. Their howlings and cla-
mours in the night are dreadful, and fo loud that people can
fcarcely hear one another fpeak. *Dellon* fays, their voice is like
the cries of a great many children of different ages mixed together: when one begins to howl, the whole pack join in the cry.
Kæmpfer fays, that every now and then a fort of bark is intermixed; which confirms what I above affert. *Dellon* agrees in the
account of their being tamed, and entertained as domeftic animals. During day they are filent.

They dig burrows in the earth, in which they lie all day, and
come out at night to range for prey: they hunt by the nofe, and
are very quick of fcent *.

The females breed only once a year; and go with young only
four weeks: they bring from fix to eight at a time †.

Both Mr. *Gueldenftaedt* and Mr. *Bell* contradict the opinion of
their being very fierce animals.

This animal is vulgarly called the Lion's Provider, from an
opinion that it rouzes the prey for that bad-nofed quadruped.
The fact is, every creature in the foreft is fet in motion by the

* *Bell's trav.* i. 54. 55.
† *Gmelin,* jun. as quoted by Mr. *Zimmerman,* p. 473.

fearful

fearful cries of the Jackals; the Lion, and other beasts of rapine, by a sort of instinct, attend to the chace, and seize such timid animals as betake themselves to flight at the noise of this nightly pack. Described by *Oppian** under the name of Λυκ☉ Ξυθος, or *yellow wolf*; who mentions its horrible howl.

May, as M. *de Buffon* conjectures, be the Θως of *Aristotle*†, who mentions it with the wolf, and says that it has the same (I suppose partial) internal structure as the wolf, which is common with congenerous animals. The *Thoes* of *Pliny* may also be a variety of the same animal; for his account of it agrees with the modern history of the *Schakal*, except in the last article ‡.

Capische Schacall. *Schreber Germ.* ii. *tab.* xcv. *p.* 370, & Canis Mesomelas.   173. CAPESCH.
      The same. *Tenlie* or *Kenlie* of the *Hottentots*.

D. with erect yellowish brown ears, mixed with a few scattered black hairs: head of a yellowish brown, mixed with black and white, growing darker towards the hind part: sides of a light brown, varied with dusky hairs: of the body, and also the back part of the legs, of a yellowish brown, lightest on the body: throat, breast, and belly, white.

On the neck, shoulders, and back, is a bed of black; broad on the shoulders, and growing narrower to the tail, where the hairs

* *Cyneg.* iii. 296.
† *Hist. An. lib.* ii. *c.* 17. *lib.* ix. *c.* 44.
‡ *Thoes*, Luporum id genus est procerius longitudine, brevitate crurum dissimile, velox saltu, venatu vivens, *innocuum homini*. Lib. viii. *c.* 34.

are fmooth.   The part on the neck feems barred with white: that on the fhoulders with white conoid marks, one within the other, the end pointing to the back: when the hairs are ruffled, thefe marks vanifh, or grow lefs diftinct, and a hoarinefs appears in their ftead.

The tail is bufhy, of a yellowifh brown: marked on the upper part with a longitudinal ftripe of black, and towards the end encircled with two rings of black, and is tipt with white.

SIZE.   Length two feet three quarters, to the origin of the tail: the tail one foot.

PLACE.   Inhabits the countries about the *Cape of Good Hope*, and probably is found as high as the Line.

174. CEYLONESE.   Chien fauvage de Ceylan. *Vofmaer*.

D. with a long thick nofe, blunt at the end: ears erect at their bottom, pointing forward at their ends: the legs ftrong: the claws more like thofe of a cat than a dog: the color cinereous yellow: belly afh-colored: the legs almoft entirely brown: the hair clofe-fet, and foft.

SIZE.   The length of the body twenty-two inches and a half, of the tail fixteen.   The tail tapers to a point.

PLACE.   This animal is a native of *Ceylon*: its hiftory quite unknown.

Canis

Ceylonese Dog. — N.º 74.

Zerda. N.º 176.

Canis Thous. **C.** cauda deflexa lævi, corpore fubgrifeo fubtus albo. *Lin. fyft.* 60. 175. SURINAM.

D. with upright ears: little warts on the cheeks, above the eyes, and under the throat: the tongue fringed on the fides: fize of a large cat: color of the upper part of the body greyifh; the lower white: tail bending downwards, and fmooth: five toes before, four behind.

According to *Linnæus*, inhabits *Surinam*: mentioned by no other Naturalift.

Stock. Wettfk. Handl. 1777, *p.* 265, *tab.* vi. 176. ZERDA.

D. with a very pointed vifage: long whifkers: large bright black eyes: very large ears, of a bright rofe-color; internally lined with long hairs: the orifice fo fmall as not to be vifible, probably covered with a valve or membrane: legs and feet like thofe of a dog: tail taper.

Color between a ftraw and pale brown. COLOR.

Length, from nofe to tail, ten inches: ears three inches and a half long: tail fix: height not five. SIZE.

Inhabits the vaft defert of *Saara*, which extends beyond Mount *Atlas*: is called by the *Moors*, *Zerda*: burrows in the fandy ground, which fhews the ufe of the valves to the ears: is fo exceffively fwift, that it is very rarely taken alive: feeds on infects, efpecially locufts: fits on its rump: is very vigilant: PLACE.

M m 2   barks

barks like a dog, but much shriller, and that chiefly in the night: never is observed to be sportive. Doctor *Sparman* suspects that he saw it during his travels in *Caffraria* *

We are indebted to Mr. *Eric Skioldebrand*, the late *Swedish* Consul at *Algiers*, for our knowledge of this singular animal. He never could procure but one alive, which escaped before he examined its teeth: the genus is very uncertain: the form of its head and legs, and some of its manners, determined us to place it here. That which was in possession of Mr. *Skioldebrand* fed freely from the hand, and would eat bread or boiled meat. Mr. *Skioldebrand* had a drawing made of the animal, and we are informed that he communicated a copy of it to Mr. *Bruce*, at that time the *British* consul at *Algiers*. This is a secret betrayed by Doctor *Sparman*, which brings on him the wrath of Mr. *Bruce*, expressed in terms I cannot repeat †. Mr. *Bruce* claimes the honor of the drawings, and asserts, that Mr. *Skioldebrand* acquired the copy by unfair means; that he corrupted his servant, and gained his end. This never would have been known, but by the lucky accident of a death-bed repentance: the poor lad fell ill; nor could he depart in peace till he had discharged his conscience by a full confession of his grievous crime. The world will probably think,

      Nec Deus intersit, nisi dignus vindice nodus
      Inciderit.

M. *de Buffon* ‡ has given a figure of this animal, communicated to him by Mr. *Bruce*; but from his authority ascribes to it a different place, and different manners. He says that it is found to the

---

    * Vol. ii. p. 186.   † Mr. *Bruce's Travels*, v. 129.
         ‡ *Supplem*. iii. 48. *tab*. xix.

fouth of the *Palus Tritonides*, in *Libya*; that it has fomething of the nature of the hare, and fomething of the fquirrel; and that it lives on the palm-trees, and feeds on the fruits.

When Mr. *Bruce* favored the public with his fplendid work, he gives at p. 128 of his fifth volume a different account. From a hare or a fquirrel, it is converted into a weefel; and the place of its habitation is changed from the *Palus Tritonides* to *Bifcara*, a fouthern province of *Mauritania Cæfarienfis*, many hundred miles from the firft pofition.

I will not dare to fix any genus to this curious and feemingly anomalous animal. To judge by Mr. *Bruce's*, or Mr. *Skiolde-brand's* figure (I will not attempt to decide the property), it has all the appearance of the *vulpine*: its face ftrongly fhews the alliance; and the length and ftrength of limbs are other proofs, very fatisfactory proofs, of its being no more able, with limbs fo formed, to climb a tree, than a dog. All the weefel tribe have very fhort legs: they can climb; they do creep. Our great RAY makes the laft the character of the clafs, and for that reafon ftyles them *vermineum* genus, the *vermes*, or worm-like clafs. Had the figure received that form of limb, I would have affented to the genus, nor even have troubled the public or myfelf, with my difference of opinion with the great traveller.

Six

XVIII. HYÆNA. Six cutting teeth, and two canine, in each jaw,
Four toes on each foot.
Short tail; a tranſverſe orifice between it and the anus.

177. STRIPED. Ϊαινα. *Ariſt. hiſt. an. lib.* vi. *c.* 32. *Op-pian Cyneg.* iii. 263.
Hyæna. *Plinii lib.* viii. *c.* 30.
Lupus marinus. *Belon aquat.* 33. *Geſner quad.*
Taxus porcinus, ſive Hyæna veterum. *Kaſtoar, Kæmpfer Amœn. exot.* 411.
Dubha. *Shaw's travels,* 246.

Hyæna. *Ruſſel's Aleppo,* 59.
Canis Hyæna. C. cauda recta annulata, pilis cervicis erectis, auriculis nudis, palmis tetradactylis. *Lin. ſyſt.* 58.
L'Hyæne. *De Buffon,* ix. 268 *tab.* xxv. *Briſſon quad.* 169. *Schreber,* xcvi.

H. with long ſharp-pointed naked ears: upright mane: high ſhoulders: fore legs longer than the hind legs: hair on the body coarſe, rough, and pretty long: of an aſh-color, marked with long black ſtripes from the back downwards: others croſs the legs: tail very full of hair, ſometimes plain, ſometimes barred with black: ſize of a large dog, but very ſtrongly made.

Inhabits the mountains of *Caucaſus* and the *Altaic Chain, Aſiatic Turky, Syria, Perſia, Barbary,* and *Senegal,* and even as low as the *Cape* *. Is by *Adanſon* and others frequently miſnamed, the *wolf,* which is not even found in *Africa:* like the jackal, violates the repoſitories of the dead, and greedily devours the putrid contents of the grave; preys by night on the herds and flocks; yet, for want of other food, will eat the roots of plants†, and the tender ſhoots of the palms; but, contrary to the nature of the former, is

* *Forſter.*　　† *Shaw's Travels,* 246.
3　　　　　　　　　　　　　　　　　　　an

an unfociable animal; is folitary, and inhabits the chafms of the rocks: will venture near towns; and, as Mr. *Niebuhr* affures us, will, about *Gambron*, in the feafon when the inhabitants fleep in the open air, fnatch away children from the fides of their parents *. The fuperftitious *Arabs*, when they kill one, carefully bury the head †, leaft it fhould be applied to magical purpofes; as the neck was of old by the *Theffalian* forcerefs.

> *Vifcera non Lyncis, non diræ nodus* Hyænæ
> *Defuit* ‡.

> Nor entrails of the fpotted *Lynx* fhe lacks,
> Nor bony joints from fell *Hyæna*'s backs.

ROWE.

The antients were wild in their opinion of the *Hyæna*: they believed that it changed its fex; imitated the human voice; that it had the power of charming the fhepherds, and as it were riveting them to the place they ftood on: no wonder that an ignorant *Arab* fhould attribute to its remains preternatural powers.

They ufually are cruel, fierce, and untameable animals, with a moft malevolent afpect: have a fort of obftinate courage, which will make them face ftronger quadrupeds than themfelves; *Kæmpfer* relates that he faw one which had put two lions to flight. Their voice is hoarfe, a difagreeable mixture of growling and roaring.

* *Defcr. Arabie,* 147.
† *Shaw's Travels,* 246.
‡ *Lucan, lib.* vi. 672. The antients believed that the neck of the *Hyæna* confifted of one bone without any joint.

I recollect

I recollect an instance, an exception to the notion of their untameable nature; having feen one at Mr. *Brook*'s as tame as a dog.
M. *de Buffon* mentions another: it is probable that if they are
taken very young, they may be reclamed by good ufage; but
they are commonly kept in a perpetual ftate of ill humor by the
provocations of their mafter.   I faw this year (1792) in the Tower
two young ones not above half a year old.   They were quite tame
and inoffenfive: but I was informed that, as they advanced in
life, their favage nature would appear.

178. SPOTTED.   Jackal, *or* Wild Dog. *Bofman's Guinea,*   Hyæna, *or* Crocuta? *Ludolph. Æthiopia,*
                   293.   57.
                  Quumbengo. *Churchill's coll. voy.* v. 486.   Cani-apro-lupo-vulpes? *Deflandes Hift.*
                  Tiger Wolf. *Kolben's Cape,* ii. 108.   *de l'Acad. tom.* xxviii. 50. *octavo ed.*

H. with a large and flat head: above each eye fome long hairs:
on each fide of the nofe very long whifkers: fhort black
mane: hair on the body fhort and fmooth: ears fhort, and a little
pointed; their outfide black, infide cinereous: face, and upper
part of the head, black: body and limbs reddifh brown, marked
with diftinct round black fpots; the hind legs with tranfverfe
black bars: tail fhort, black, and full of hair.   This defcription
was taken from one fhewn fome years ago in *London.*   It was fuperior in fize to the former.

Inhabits *Guinea, Æthiopia,* and the *Cape:* lives in holes in the
earth, or clefts of rocks: preys by night: howls horribly: breaks
into the folds, and kills two or three fheep: devours as much as
it can, and carries away one for a future repaft: will attack mankind; fcrape open graves, and devour the dead.   It has very great
                                                              ftrength

Spotted Hyæna____. No 178.

ftrength. One has been known to feize a female *Negro*, fling her
over its back, and holding her by one leg, run away with her till
fhe was fortunately refcued*. M. *de Buffon*, mifled by *Bofman*'s
name of this animal, makes it fynonymous with the common
jackal. Has, till the prefent time, been undiftinguifhed by naturalifts.

M. *de Buffon* had an account from Mr. *Bruce*, of an *Hyæna*
which that gentleman obferved in the ifle of *Meroe*, in *Æthiopia*.
He fays that it was greatly fuperior in fize to the common kind;
had a head more like that of a dog, and a very wide mouth; without
a mane on the neck; perhaps it was not obferved, on account of
its fhortnefs. He adds this proof of its ftrength, that it would
lay hold of a man, lift him up with the greateft eafe, and run a
league or two with him, without onee putting him on the
ground †. Can there be any doubt but that the traveller meant
the fame animal with this?

* * *Bofman*, 295.   † *De Buffon, Supplem.* iii. 235.

XIX. CAT.　　　　Six cutting teeth, and two canine, in each jaw.
　　　　　　　　　Five toes before; four behind.
　　　　　　　　　Sharp hooked claws, lodged in a sheath, that may be ex-
　　　　　　　　　　　erted or drawn in at pleasure.
　　　　　　　　　Round head, and short visage: rough tongue.

*With long tails.

179. LION.　　Leo. *Plinii lib.* viii. *c.* 16. *Gesner quad.*　　Felis Leo.　F. cauda elongata, corpore
　　　　　　572. *Raii syn. quad.* 162.　　　　　　　helvulo. *Lin. syst.* 60.
　　　　　　Lowe. *Klein quad.* 81.　　　　　　　　　Le Lion.　*De Buffon,* ix. 1. *tab.* i. ii.
　　　　　　Felis cauda in floccum desinente. *Bris-*　　Lev. Mus.
　　　　　　son quad. 194. *Schreber,* xcvii. A. B.

C. with a large head: short rounded ears: face covered with
short hairs; upper part of the head, chin, whole neck, and
shoulders, with long shaggy hairs, like a mane: hair on the body
and limbs short and smooth; along the bottom of the belly long:
limbs of vast strength: tail long, with a tuft of long hairs at the
end: color tawny, but on the belly inclines to white: length of
the largest lion, from nose to tail, above eight feet: the tail four
feet long; tufted with long black hairs: the lioness or female is
less, and wants the mane.

An inhabitant of most parts of *Africa*; and rarely of the hot
parts of *Asia,* such as *India* * and *Persia*†; and a few are still met

Fryer's voy. 189. *Bernier*'s voy. *Kachemir,* 48.
† In *Gilan* and *Cardistan.* See the new description of *Persia* in *Harris*'s *Coll.*
ii. 884.

　　　　　　　　　　　　　　　　　　　　　　with

with in the deserts between *Bagdat* and *Bassorah* [*], on the banks
of the *Euphrates*. Mr. *Niebuhr* also places them among the animals of *Arabia* [†]; but their proper country is *Africa*, where their
size is the largest, their numbers greatest, and their rage more
tremendous, being inflamed by the influence of a burning sun, on
a most arid soil. Doctor *Fryer* says, that those of *India* are feeble
and cowardly. In the interior parts [‡], amidst the scorched and
desolate deserts of *Zaara*, or *Biledulgerid*, they reign sole masters;
they lord it over every beast, and their courage never meets with
a check, where the climate keeps mankind at a distance: the
nearer they approach the inhabitants of the human race, the less
their rage, or rather the greater is their timidity [||]: they have
often experienced the unequal combat, and, finding that there
exists a being superior to them, commit their ravages with more
caution: a cooler climate again has the same effect; but in the
burning deserts, where rivers and fountains are denied, they live
in a perpetual fever, a sort of madness fatal to every animal they
meet with. The author of the *Oeconomy of Nature* gives a won-
derful proof of the instinct of these animals in those unwatered
tracts. There the *Pelican* makes her nest; and in order to cool
her young ones, and accustom them to an element they must afterwards be conversant in, brings from afar, in their great gular
pouch, sufficient water to fill the nest; the lion, and other wild
beasts, approach and quench their thirst, yet never injure the
unfledged [§] birds, as if conscious that their destruction would

---

[*] Voyages *de Boullaye Le Gouz*, 320.   [†] *Descr. Arabie*, 142.
[‡] *Leo Afr.* 342.   [||] *Purchas's Pilg.* ii. 809,   [§] *Amœn. Acad.* ii. 37.

immediately

immediately put a ftop to thofe grateful fupplies. It is to be obferved, that whenever a lion can get at water, it drinks much.

The courage of the lion is tempered with mercy*, and has been known to fpare the weaker animals, as if beneath his attention: there are many inftances of its gratitude; relations fo ftrange, that the reader is referred to them in the notes † to the authorities themfelves. Lions are capable of being tamed: the monarch of *Perfia*, full of favage ftate, had, on days of audience ‡, two great ones chained on each fide of the paffages to the room of ftate, led there by keepers, in chains of gold. As they have been fo far fubdued, why may we not credit the ftory of their being harneffed for the triumphal car of the conqueror *Bacchus?*

The lion preys on all kinds of animals: as his fcent is bad, his peculiar and tremendous roar ftrikes terror into every beaft of the defert, and fets them in motion, in open view; he then felects his object, and takes it not fo much by purfuit, as by a vaft bound, ftriking it with his talons, and tearing it to pieces. In inhabited countries he invades the folds, leaps over the fences with his prey; and fuch is his ftrength, that he can carry off a middling ox with the utmoft eafe ‖: in many places it takes its prey by furprize, lurking in the thickets, and fpringing on it: oft-times mankind fall a victim to his hunger, but then it is rather thro' neceffity than choice. The *Arabs* have a notion of his fparing the tender fex; but Doctor *Shaw* informs us § that they

---

* *Leoni tantum ex feris clementia in fupplices: proftratis parcit: et ubi fævit, in viros prius, quam in fæminas fremit, in infantes non nifi magna fame.* Plinii lib. viii. c. 16. Miffon, vol. iii. 292, *confirms the laft.*

† *A. Gellius. Ælian. Pliny.*   ‡ *Bell's travels,* i. 102.   ‖ *La Caille,* 294. § *Travels,* 244.

make

make no diftinction in thefe days : the fame writer acquaints us, that the flefh of the lion is often eaten in *Barbary*, and it refembles veal in tafte.

Formerly found in *Europe*, between the rivers *Achelous* and *Neffus* *; none in *America*; the animal called *Puma* †, which is miftaken for the lion, is our 160th fpecies.

Tigris. *Plinii lib.* viii. *c.* 18.   *Bontii*   *fyft.* 61.   180. TIGER.
  Java, 5:. *Gefner quad.* 936.   *Raii*   Felis flava, maculis longis nigris varie-
  fyn. quad. 165. *Klein quad.* 78.   gata. *Briffon quad.* 194.
Felis Tigris. F. cauda elongata, cor-   Le l'igre. *De Buffon*, ix. 129. *tab.* ix.
  pore maculis omnibus virgatis. *Lin.*   *Schreber*, xcviii. LEV. MUS.

C. with a fmooth head and body; vaft ftrength in its limbs; of a pale yellow color, beautifully marked with long ftripes of black from the back, pointing to the belly, with others crofs the thighs: the tail fhorter by a third than the body; annulated with black: often fuperior in fize to a lion; that called the *Royal* ‡ Tiger is of a tremendous bulk. M. *de Buffon* mentions one that was (tail included) fifteen· feet long. *Hyder Ally* prefented the Nabob of *Arcot* with one of far greater dimenfions, it being eighteen feet in length. *Du Halde*, ii. 254, fays, that the *Chinefe* tiger, or *Lou-chu*, or *Lau-hu*, as it is called in that language, varies in color, fome being white, ftriped with black and grey.

* *Ariftot. hift. an. lib.* vi. *c.* 31.
† *Garcilaffo de la vega,* 332.
‡ *Dellon voy.* 78.

The

C A T.

The tiger is peculiar to *Asia* *; and is found as far north as *China*, and *Chinese Tartary*; and about lake *Aral*, and the *Altaic* mountains. By a moſt common miſnomer, this animal is improperly given to *Africa* and *America*. It inhabits mount *Ararat*, and *Hyrcania*, of old famous for its wild beaſts; but the greateſt numbers, the largeſt, and the moſt cruel, are met with in *India*, and its iſlands. In *Sumatra* the natives are ſo infatuated that they ſeldom kill them, having a notion that they are animated by the ſouls of their anceſtors †. They are the ſcourge of the country; they lurk among the buſhes, on the ſides of rivers, and almoſt depopulate many places: they are inſidious, blood-thirſty, and malevolent; and ſeem to prefer preying on the human race preferable to any other animals: they do not purſue their prey, but bound on it from their ambuſh, with an elaſticity, and from a diſtance that is ſcarcely credible: if they miſs the object, they make off; but if they ſucceed, be it man or be it beaſt, even one as large as a *Buffalo* ‡, they carry it off with ſuch eaſe, that it ſeems not the leſt impediment to their flight. If they are undiſturbed, they plunge their head into the body of the animal up to their very eyes, as if it were to ſatiate themſelves with blood, which they exhauſt the corpſe of before they tear it to pieces ‖. There is a ſort of cruelty

MANNERS.

* M. *de Buffon* ſays they are found in the ſouth of *Africa*. I can meet with no authority for it; the animals ſo called by *Ludolphus* and *Kolben*, being only Panthers, or Leopards, which are generally confounded with the Tiger by moſt voyagers.

† Mr. *Miller's* Account of *Sumatra*, Phil. Tranſ. lxviii. 171.

‡ *Bontius*, 53. *Strabo. lib.* xv. relates much the ſame of the Tigers of the country of the *Praſii*.

‖ *Bontius*, 53.

in

ın their devaſtations, unknown to the generous lion; as well as a poltronery in their ſudden retreat on any diſappointment. I was informed, by very good authority, that in the beginning of this century, ſome gentlemen and ladies, being on a party of pleaſure, under a ſhade of trees, on the banks of a river in *Bengal*, obſerved a tiger preparing for its fatal ſpring; one of the ladies, with amazing preſence of mind, laid hold of an umbrella, and furled it full in the animal's face, which inſtantly retired, and gave the company opportunity of removing from ſo terrible a neighbor.

Another party had not the ſame good fortune: a tiger darted among them while they were at dinner, ſeized on one gentleman, and carried him off, and he never was more heard of. They attack all ſorts of animals, even the lion; and it has been known that both have periſhed in their combats. There is in ſome parts of *India* a popular notion *, that the rhinoceros and the tiger are in friendſhip, becauſe they are often found near each other: the faſt is, the rhinoceros, like the hog, loves to wallow in the mire; and on that account frequents the banks of rivers; the tiger, to quench its raging thirſt, is met with in places contiguous to them.

Pliny has been frequently taken to taſk by the moderns, for calling the tiger, *animal tremendæ velocitatis* †; they allow it great agility in its bounds, but deny it ſwiftneſs in purſuit: two travellers of authority, both eye-witneſſes, confirm what *Pliny* ſays; the one indeed only mentions in general its vaſt fleetneſs; the

GREAT SWIFT-NESS.

* *Bontius*, 53. † *Plinii lib.* viii. *c.* 18.

other

other faw a tryal between one and a fwift horfe, whofe rider efcaped merely by getting in time amidft a circle of armed men. The chace of this animal was a favorite diverfion with the great CAM-HI, the *Chinefe* monarch, in whofe company our country-man, Mr. *Bell*, that faithful traveller, and the *Pere Gerbillon*, faw thefe proofs of the tiger's fpeed *.

They are faid to roar like a lion; but thofe I have feen in cap-tivity, emitted only a furly growl.

181. PANTHER.

Varia et Pardus? *Plinii lib.* viii. *c.* 17. Παρδαλις μειζων? *Oppian Cyneg. lib.* iii. *l.* 63.
Panthera, Pardus, Pardalis, Leopar-dus. *Gefner quad.* 824. *Raii fyn. quad.* 166. *Klein quad.* 77.

Felis Pardus. F. cauda elongata, cor-pore maculis fuperioribus orbicula-tis; inferioribus virgatis. *Lin. fyft.* 61 †. *Briffon quad.* 198.
La Panthere. *De Buffon*, ix. 151. *tab.* xi. xii. *Schreber*, xcix.

C with fhort fmooth hair, of a bright tawny color: the back, fides, and flanks, elegantly marked with black fpots, dif-pofed in circles, from four to five in each, with a fingle black fpot in the centre of each: on the face and legs fingle fpots only: on the top of the back is a row of oblong fpots; the longeft next the tail: the cheft and belly white; the firft marked with tranf-verfe dufky ftripes: the belly and tail with large irregular black fpots: ears fhort and pointed: end of the nofe brown: limbs very ftrong: the fkin of one I meafured was, from the end of the nofe to the origin of the tail, fix feet ten inches; the tail near three.

* *Bell's Travels,* ii. 91. *Du Halde,* ii. 343.
† A defcription that does not fuit any known animal of this genus.

Inhabits

Inhabits *Africa*, from *Barbary* to the remoteft parts of *Guinea* *. This fpecies is next in fize to the tiger; next to it in cruelty, and in its general enmity to the animal creation: it is to *Africa* what the former is to *Afia*, with this alleviation, that it prefers the flefh of brutes to that of mankind; but when preffed with hunger, attacks every living creature without diftinction: its manner of taking its prey is the fame with that of the tiger, always by furprize, either lurking in thickets, or creeping on its belly till it comes within reach: it will alfo climb up trees in purfuit of monkies, and leffer animals: fo that nothing is fecure from its attacks. It is an untameable fpecies; always retains its fierce, its malevolent afpect, and perpetual growl or murmur.

The antients were well acquainted with thefe animals; thefe and the leopards were the *Variæ* and *Pardi* of the old writers: one fhould think that the *Romans* would have exhaufted the deferts of *Africa*, by the numbers they drew from thence for their public fhews: *Scaurus* exhibited at one time 150 *Panthers*; *Pompey* the Great 410; *Auguftus* 420†: probably they thinned the coafts of *Mauritania* of thefe animals, but they ftill fwarm in the fouthern parts of *Guinea*. This fpecies, the *Leopard*, and the *Once*, were obferved by Doctor *Sparman* as remote as the Cape of *Good Hope*‡.

In my former edition I ufed fome arguments in favor of thefe animals being alfo natives of *South America*. I had feen the fkins at the furriers fhops, which had been brought from the *Brazils*: but as that country has a great intercourfe with *Congo* and *Angola*

---

* *Shaw's Travels*, 244. *Des Marchais*, i. 204. the laft miftakenly calls them Tigers.

† *Plinii lib.* viii. *c.* 17.   ‡ *Travels.* ii. 251.

on account of the Slave Trade, I have no doubt but that they were brought from thofe kingdoms, and re-exported to *Europe.* The largeft congenerous animal that *South America* has is the *Brafilian,* hereafter to be tranfcribed.

Oppian defcribes two fpecies of *Panthers;* a large fpecies and a fmall one; the firft of which has a fhorter tail than the leffer, and may poffibly be this kind.

182. LEOPARD. Uncia. *Caii opufc.* 42. *Gefner quad.* 825. Le Leopard. *De Buffon,* ix. 151. *tab.* xiv. Le Leopard. *De Marcha.s voy.* i. 202. *Schreber,* ci. LEV. MUS.

C with hair of a lively yellow color; marked on the back and fides with fmall fpots, difpofed in circles, and placed pretty clofely together: the face and legs marked with fingle fpots: the breaft and belly covered with longer hairs than the reft of the body, of a whitifh color: the fpots on the tail large and oblong: the length of this fpecies, from nofe to tail, four feet; the tail two and a half.

Inhabits *Senegal* and *Guinea;* fpares neither man nor beaft: when beafts of chace fail, defcends from the internal parts of *Africa* in crowds, and makes great havock among the numerous herds that cover the rich meadows of the lower *Guinea.* It tears its prey to pieces with both claws and teeth; is always thin, tho' perpetually devouring. The Panther is its enemy, and deftroys numbers of them. The Negreffes make collars of their teeth, and attribute to them certain virtues. The Negroes take thefe animals in pit-falls, covered at the top with flight hurdles, on

which

Black_Leopard.

which is placed fome flefh as a bait. The Negroes make a banquet of thefe animals, whofe flefh is faid to be as white as veal, and very well tafted. The fkins are often brought to *Europe*, and reckoned very valuable.

In *Afia* it is found in the mountains of *Caucafus*, from *Perfia* to *India*; and alfo in *China*, where it is called *Poupi*; and by the *Bucharian* traders, who often bring their fkins to *Ruffia*, are ftyled *Bars*. It inhabits alfo *Arabia*, where it is called *Nemr*. We are informed by Mr. *Forfkal**, that in that country, as well as in *Ægypt*, it will do no harm to man unlefs provoked; but will enter houfes by night, and deftroy the cats.

In the Tower of *London* is a black variety, brought from *Bengal* by *Warren Haftings*, Efq. The color univerfally is a dufky black, fprinkled over with fpots of a gloffy black, difpofed in the fame forms as thofe of the Leopard: on turning afide the hair, beneath appears a tinge of the natural color. BLACK VARIETY.

This animal is engraven by M. *De la Metherie*†. That gentleman mentions my quoting the *Congar noire* of M. *de Buffon* as a fynonym. I beg leave to rectify his miftake. The black *Tiger* is a diftinct fpecies, and from a different country, being a native of *South America*. I muft fay befides, that M. *de Buffon* was totally unacquainted with the animal till I fent to him the drawing from which he made the engraving in vol. iii. of his fupplement, tab. xlii. notwithftanding he fuppreffes the origin.

* P. v.
† Obfervations fur la Phyfique, &c. tom. xxxviii. *Juillet.* 1788. p. 45.
O o 2   C. with

183. LESSER
LEOPARD.

C. with the face fpotted with black: chin white: a great black fpot on each fide of the upper lip: breaft marked with fmall fpots: belly white, fpotted with black: back, fides, and rump, covered with hair of a bright yellow color: marked with circles of fpots, like the former; but the fpots much lefs: not half the bulk of the laft; but the tail fhorter in proportion, and tapering to a point, and the hair on it fhort. The tails of the two laft fpecies are of equal thicknefs from top to bottom.

Inhabits the *Eaft Indies?* kept a few years ago in the Tower: feemed a good-natured animal.

184. HUNTING.

Le Leopard. *Voy. de Boullaye-le-gouz,* 248.
Felis jubata. *Schreber,* cv.

Le Gueparp. *De Buffon,* xiii. 249.
Le Jaguar, ou le Leopard. *Suppl.* 218.
tab. xxxviii. LEV. MUS.

C. with a fmall head: irides pale orange: end of the nofe black: from each corner of the mouth to that of each eye, a dufky line: ears fhort, tawny, marked with a brown bar: face, chin, and throat, of a pale yellowifh brown: the face flightly fpotted: body of a light tawny brown, marked with numbers of fmall round black fpots; not in circles, but each diftinct: the fpots on the rim and outfide of the legs were larger: the infide of the legs plain: hair on the top of the neck longer than the reft: that on the belly white, and very long: tail longer than the body; of a reddifh brown color; marked above with large black fpots; the hair on the under fide very long.

3

Size

Hunting Leopard.___ N.º 284.

Size of a large gre-hound: of a long make: cheſt narrow: legs very long.

Inhabits *India:* is tamed, and trained for the chace of ante-lopes: carried in a ſmall kind of waggon, chained and hood-winked, till it approaches the herd: when firſt unchained, does not immediately make its attempt, but winds along the ground, ſtop-ping and concealing itſelf till it gets a proper advantage, then darts on the animals with ſurprizing ſwiftneſs: overtakes them by the rapidity of its bounds: but if it does not ſucceed in its firſt efforts, conſiſting of five or ſix amazing leaps, it miſſes its prey: loſing its breath, and finding itſelf unequal in ſpeed, ſtands ſtill; gives up the point for that time*, and readily returns to its maſter.

This ſpecies is called in *India,* *Chittah.* It is uſed for the tak-ing of jackals, as well as other animals.

Παρδαλις. *Oppian Cyneg.* iii. *l.* 95. L'Once. *De Buffon,* ix. 151. *tab.* xiii. 185. ONCE,
Panthera? *Plinii lib.* viii. *c.* 17. *Schreber,* c.

C. with a large head: ſhort ears: long hair on the whole bo-dy: color a whitiſh aſh, tinged with yellow; on the breaſt and belly with a ſmaller caſt of yellow: head marked with ſmall round ſpots: behind each ear a large black ſpot: the upper part of the neck varied with large ſingle ſpots: the ſides of the back with longitudinal marks, conſiſting of ſeveral ſpots, almoſt touch-ing each other, leaving the ground color of the body in the mid-dle: the ſpots beneath theſe irregular, large, and full: thoſe on the legs ſmall, and thinly diſperſed: the tail full of hair; irregu-

* *Bernier's travels,* iv. 45. *Tavernier's travels,* i. 147. *Thevenot, voy.* v. 34.

larly

larly marked with large black fpots. This fpecies is of a ft.ong make: long backed: fhort legged: length, from the nofe to the tail, about three feet and a half: tail upwards of three feet.

Inhabits *Barbary* *, *Perfi.1*, *Hyrcania* †, and *China* ‡; the *Bucharian* and *Altaic* chain, and to the weft of Lake *Baikil*: is an animal of a more gentle and mild nature than moft of the preceding; is, like the laft, ufed for the chace of antelopes, and even hares; but, inftead of being conveyed in a waggon, is carried on the crupper on horfeback: is under as much command as a fetting-dog; returns at the left call, and jumps up behind its mafter ‖.

Is fuppofed to be the leffer Panther of *Oppian*, and the *Panthera* of *Pliny* §.

---

186. BRASILIAN.

Jagura. *Marcgrave Brafil.* 235. *Pifo Brafil.* 203.
Pardus aut Lynx *Brafilienfis*, Jaguara dicta, *Lufitanis* onza. *Raii fyn. quad.* 168. *Klein quad.* 80.
Le Tigre de la Guiane. *Des Marchais, voy.* iii. 299.
Tigris Americana. Felis flavefcens, maculis nigris orbiculatis quibufdam rofam referentibus variegata. *Briffon quad.* 196.
Felis onça. Felis cauda mediocri, corpore flavefcente, ocellis nigris rotundato angulatis medio flavis. *Lin. fyft.* 91.
Le Jaguar. *De Buffon*, ix. 201. tab. xviii. *Suppl.* iii. 218. *tab.* xxxix. *Schreber*, cii.

C with hair of a bright tawny color: the top of the back marked with long ftripes of black: the fides with rows of ir-

---

* Where it is called *Faadh.* *Shaw's trav.* 245.
† *Chardin.*
‡ The fkins are brought from *China* into *Ruffia*, and fold for twenty fhillings a piece. *Muller Samlunge rur Ruffifchen Gefchicht.* iii. 549, 608.
‖ *Olearius's travels into Perfia*, 218.
§ Pantheris *in candido breves macularum oculi*, lib. viii. c. 17.

regular

Brasilean Tiger.___ N° 186.

regular oblong fpots; open in the middle, which is of the ground-color of the hair: the thighs and legs marked with full fpots of black: the breaft and belly whitifh: the tail not fo long as the body; the upper part deep tawny, marked with large black fpots, irregularly; the lower part with fmaller fpots: grows to the fize of a wolf, and even larger.

Inhabits the hotteft parts of *S. America*, from the ifthmus of *Darien* to *Buenos Ayres*: fierce and deftructive to man and beaft. Like the tiger, it plunges its head into the body of its prey, and fucks out the blood before it devours it: makes a great noife in the night, like the howling of a hungry dog: is a very cowardly animal: eafily put to flight; either by the fhepherds dogs, or by a lighted torch, being very fearful of fire: it lies in ambufh near the fides of rivers: there is fometimes feen a fingular com-bat between this animal and the crocodile; when the *Jaguar* comes to drink, the crocodile, ready to furprize any animal that approaches, raifes its head out of the water, the former inftantly ftrikes its claws into the eyes of this dreadful reptile, the only penetrable part, who immediately dives under the water, pulling his enemy along with it, where they commonly both perifh *.

Tlacoozelotl; Tlalocelotl. Catul-par-dus Mexicanus. *Hernandez Mex.* 512. L'Ocelot. *De Buffon*, xiii. 239. *tab.* xxxv. xxxvi. Felis Pardalis. *Lin. fyft.*

Felis fylveftris, Americanus, Tigrinus. *Seb. Muf.* i. 47. *tab.* xxx. *fig.* 2, & 77. *tab.* xlviii. *fig.* 2. *Schreber*, ciii.

187. MEXICAN.

C. with its head, back, upper part of the rump, and tail, of a bright tawny: a black ftripe extends along the top of the

* *Condamine's voy.* 81.

back,

back, from head to tail: from the noftrils to the corners of the eyes, a ftripe of black: forehead fpotted with black: the fides whitifh, marked lengthways with long ftripes of black, hollow and tawny in the middle; in which are fprinkled fome fmall black fpots: from the neck towards the fhoulders, point others of the fame colors: the rump marked in the fame manner: legs whitifh, varied with fmall black fpots: tail fpotted with fmall fpots near its bafe; with larger near the end, which is black: about four times the fize of a large cat.

PLACE.   Inhabits *Mexico*, the neighborhood of *Carthagena*, and *Brafil*: lives in the mountains: is very voracious; but fearful of man-kind: preys on young calves*, and different forts of game: lurks amidft the leaves of trees; and fometimes will extend itfelf along the boughs, as if dead, 'till the monkies, tempted by their natu-ral curiofity, approaching to examine it, become its prey†.

288. CINEREOUS.   C. of a cinereous color, paleft on the legs and belly: irides ha-zel: tip of the nofe red: ears fhort, and rounded; black on the outfide, grey within: from the nofe to the eye, on each fide was a black line; above and beneath each eye a white one: fides of the mouth white, marked with four rows of fmall black fpots: from the hind part of the head, to the back and fhoulders, ran fome long, narrow, hollow ftripes: along the top of the back two rows of oval black fpots: the marks on the fides long, hol-low, and irregular, extending from fhoulders to thighs: fhoul-

* *Dampier, voy.* ii. 62.
† *Hernandez, Mex.* 514.

ders

Mexican Tiger — N.° 187

ders both barred and spotted: legs and belly only spotted: tail not so long as the body; had large spots above, small beneath.

About the size of the preceding. Inhabits *Guinea*.

Cugacuarana. *Marcgrave Brasil.* 235. *Raii syn. quad.* 169.	Tigris fulva. Felis ex flavo rufescens, mento et infimo ventre albicantibus. *Brisson quad.* 197	189. Puma.
Cugacuara. *Piso Brasil.* 103.	Le Couguar. *De Buffon,* ix. 216. *tab.* xix. *Suppl.* iii. 222.	
Panther. *Lawson Carolina,* 117. *Catesby Carolina App.*		
Tigris fulvus. *Barrere France Æquin.* 166. *Du Pratz,* ii. 63.	Felis Concolor. *Schreber,* civ. Pagi. *Molina Chili.* 276.	

C. with a very small head: ears a little pointed: eyes large: chin white: back, neck, rump, sides, pale brownish red, mixed with dusky hairs: breast, belly, and inside of the legs, cinereous, hair on the belly long: tail dusky, and ferruginous; the tip black: the teeth of a vast size: claws white: the outmost claw of the fore feet much larger than the others: is long bodied, and high on its legs: the length from nose to tail five feet three inches; of the tail two feet eight.

Inhabits the continent of *America,* from *Canada* to *Brasil:* in *South America* is called *Puma**, and mistaken for the lion: is the scourge of the colonies of the hotter parts of *America*; fierce and ravenous to the highest degree: swims over the broad rivers, and attacks the cattle, even in the inclosures; and when pressed with

* *Hernandez Mex.* 518. *Condamine's voy.* 81.

hunger, spares not even mankind.   In *N. America* their fury seems to be subdued by the rigor of the climate; the smallest cur, in company with its master, makes them seek for security, by running up trees: but then they are equally destructive to domestic animals, and are the greatest nuisance the planter has: when they lay in wait for the *Moose*, or other deer *, they lie close on the branch of some tree, 'till the animal passes beneath, when they drop on them, and soon destroy them: they also make wolves their prey: that whose skin is in the *Museum* of the *Royal Society*, was killed just as it had pulled down a wolf. Conceal such part of the prey which they cannot eat: purr like a cat: their fur soft, and of some value among the *Indians*, who cover themselves with it during winter: the flesh is also eaten, and said to be as good and as white as veal†.

190. JAGUAR.   Jaguarete. *Marcgrave Brasil.* 235. *Piso*   Le Congar noir.   *De Buffon, Suppl* iii.
      Brasil. 103.   *Raii syn. quad.* 169.   223. *tab.* xlii.
      Once. *Des Marchais*, iii. 300.

C. with the head, back, sides, fore part of the legs, and the
   tail, covered with short and very glossy hairs, of a dusky-

* *Charlevoix voy. Nouv. France*, v. 189, who, by mistake, calls it *Carcajou.*

† Mr. *Dupont* once shewed me, some years ago, the tail of an animal from *South America*, three quarters of a yard long, covered with short, white, glossy hair: a piece of the skin of the back was left to it, on which were black hairs near eight inches long.   I mention it here, as belonging to some plain-colored beast of this genus; perhaps the *Tzonyztac seu quadrupes capillorum candentium, brevibus cruribus, colore atro, manibus pedibusque et corporis magnitudine Tigris; ac prolixa cauda.*   Hernandez quad. nov. Hisp. 3.

color;

Jaguar or Black Tiger. N.º 190.

color; fometimes fpotted * with black, but generally plain : upper lips white: at the corner of the mouth a black fpot†; long
hairs above each eye, and long whifkers on the upper lip: lower
lip, throat, belly, and the infide of the legs, whitifh, or very
pale afh-color : paws white : ears pointed. Grows to the fize of
a heifer of a year old: has vaft ftrength in its limbs.

Inhabits *Brafil* and *Guiana:* is a cruel and fierce beaft; much
dreaded by the *Indians;* but happily is a fcarce fpecies.

C. with fhort hair, of a bright ferruginous color: the face 191. CAPE.
marked with black ftripes, tending downwards: from the
hind part of the head to the tail, the back is marked with oblong
ftripes of black: the fides with very numerous fmall and round
fpots of black: belly white: tail long, of a bright tawny-color,
annulated with black : ears long, narrow, pointed, and very erect:
length from the nofe to the tail near three feet.

Defcribed from a fkin in a furrier's fhop in *London.* Inhabits
the neighborhood of the *Cape* of *Good Hope,* and as high north as
Congo. Inhabits the woods, and is very deftructive to lambs;
young antelopes, and all the leffer animals: is well defcribed and
figured by Doctor *Forfter,* in *Phil. Tranf.* lxxi. p. i. *tab.* i. The

---

* For which reafon *M. de Buffon* fufpects it to be only a variety of *No.* 186;
but fince M. *des Marchais,* who defcribes it very exactly, makes no mention of
its being fpotted, nor had the two which were fhewn in *London* fome years ago
any fpots on them; it is very probable, then, that the *Jaguarete,* defcribed by
Marcgrave, was a variety of this fpecies, and not of his *Jaguara,* as it agrees
with it in the ground color, and in its fuperior fize.

† On the chin of one of thofe above-mentioned was a round black fpot.

fpecimen

specimen he made his description from was only eighteen inches long. Mine might have been from a distended skin, or his from a young animal. Mr. *Miller*, in his plates, *tab.* xxxix. also gives a good figure of this animal.

---

192. CAYENNE.

Maraguao. *Marcgrave Brasil.* 233.
Felis fera tigrina. *Barrere France Æquin.* 152.
Tepe Maxlaton. *Hernand. Nov. Hisp.* 9. *c.* 28.
Le Pichou, Cat-a-mount. *Du Pratz Louisian.* ii. 64.

Felis sylvestris tigrina. F. ex griseo flavescens, maculis nigris variegata. *Brisson quad.* 193.
Le Margay. *De Buffon*, xiii. 248. *tab.* xxxvii. *Supplem.* iii. 226. *Schreber,* cvi.

C. with the upper part of the head, the neck, back, sides, shoulders, and thighs, of a bright tawny-color: the face striped downwards with black: the shoulders and body marked with stripes, and oblong large black spots: the legs with small spots: the breast, and inside of the legs and thighs whitish, spotted with black: the tail very long, marked with black, tawny, and grey: size of a common cat.

Inhabits *S. America*, and perhaps *Louisiana**; lives on the feathered game, and on poultry: is untameable: makes a noise like the common cat: lives much in trees: is very active; goes by bounds or leaps: brings forth in all seasons of the year, in hollow trees, and has two at a time.

---

193. BENGAL.

C. with white whiskers: large ears; dusky, with a white spot in the middle of the outside: between each eye and the nose a white line, and beneath each eye another.

* *Bossu's trav.* i. 94. 359.

3   Color

Color of the head, upper jaw, and fides of the neck, back, and fides, a beautiful pale yellowifh brown: the head and face ftriped downwards with black: along the back are three rows of fhort ftripes of the fame color, pointing towards the tail: behind each fhoulder, to the belly, is a black line: chin and throat white, furrounded with a femicircle of black: breaft, belly, and infide of the limbs, white; the fpots on thofe parts, the legs, and rump, round: tail long, full of hair, brown and annulated with black.

Rather lefs than a common cat, and more elegantly made.

Mr. *Lee* of *Hammerfmith*, in whofe poffeffion the remains of this animal are, affured me that it fwam on board a fhip at anchor off the coaft of *Bengal*; that after it was brought to *England*, it coupled with the female cats, which twice produced young: I faw one of the offspring, which was marked in the fame manner as the male parent; but the ground-color was cinereous. It had as little fear of water as its fire; for it would plunge into a veffel of water near two feet deep, and bring up the bit of meat flung in by way of trial. It was a far better moufer than the tame cat; and in a little time cleared Mr. *Lee*'s magazine of feeds of the fwarms of rats, which, in fpite of the domeftic breed of cats, had for a long time made moft horrible ravages among his boxes.

Thefe fmall fpotted fpecies are called by the general name of tiger cats: feveral kinds are found in the *Eaft-Indies* *, and in the woods near the *Cape of Good Hope*; but fo negligently, or fo unfcientifically mentioned, as to render it impoffible for a zoologift to form a defcription from them: yet a good hiftory of

PLACE.

MANNERS.

* *Dellon's voy.* 77.

thefe

thefe animals being among the many *defiderata* of the naturalift; the following maim accounts may ferve to direct the enquiries of future voyagers. *Kolben* * mentions two kinds; one he calls

The WILD RED CAT, which has a ftreak of bright red running along the ridge of the back to the tail, and lofing itfelf in the grey and white on the fides: the fkins are faid to give eafe in the gout, and are much valued on that account at the *Cape.* The other he calls

The BUSH CAT; of which he fays no more, than that it is the largeft of wild cats in the *Cape* countries: perhaps my *Cape* cat.

The SACA is an obfcure fpecies of wild cat, mentioned by *Flacourt* † to be found in *Madagafcar.* He fays they are very beautiful, and that they couple with the tame cats. The tails of the domeftic kind in that ifland are for the moft part turned up.

194. MANUL.　　　　　　　　Felis Manul. *Pallas Itin.* iii. App. 692.

C. with a large head: color univerfally tawny, mixed with a few white and brown hairs; crown of the head fpeckled with black: the cheeks marked with two dufky lines, running obliquely from the eyes: the feet ftriped obfcurely with dark lines: the tail longer than that of the domeftic cat, befet thickly with hair, and of an equal thicknefs in all parts; encircled with ten black rings, the three next to the tip almoft touching one another, the reft more remote.

Size of a fox: the limbs very robuft; in that and color greatly refembles a lynx.

* *Hift. Cape,* ii. 126.　　　　　† *Hift. Madag.* 152.

Inhabits

Inhabits all the middle part of northern *Afia,* from the *Yaik,* or *Ural,* as it is now called, to the very *Amur.* Loves open, woodlefs, and rocky countries, and preys on the leffer quadrupeds. Is chiefly converfant about N. Lat. 52.: for want of other retreats, it will occupy the holes of the fox or of the *Bobak.* The *Ruffians* call it *Stepnaja Kofchka,* or the *cat of the defert.*

(WILD CAT.) Catus fylveftris. Boumriitter. *Gefner quad.* 325.
Catus fylveftris, ferus vel feralis, eques arborum. *Klein quad.* 75.
Wilde Katze. *Kram Auftr.* 311.
Felis fylveftris. F. pilis ex fufco, flavicante, et albido, variegatis veftita, cauda annulis alternatim nigris et ex fordidé albo flavicantibus cincta. *Briffon quad.* 192.
Kot Driki, Zbik. *Rzaczinfki. Polon.* 217. *Schreber,* cvii. A. cvii. B.
Le chat fauvage. *De Buffon,* vi. 1. *tab.* i. *Br. Zool.* i. 67. LEV. MUS.

195. COMMON.

C. with long foft hair, of a yellowifh white color, mixed with grey; the grey difpofed in ftreaks, pointing downwards, rifing from a dufky lift, that runs from the head to tail, along the middle of the back: tail marked with alternate bars of black and white, its tip black: hind part of the legs black: three times as large as the common cat; and very ftrongly made.

Inhabits the woods of moft parts of *Europe;* but none are found in the vaft woods of *Ruffia* or *Siberia:* dwells with the common *Lynx* in all the wooded parts of the mountains of *Caucafus,* and their neighborhood: moft deftructive to lambs, kids, and fawns; and to all forts of feathered game. The ftock, or origin of the DOMESTIC CAT *, which is fubject to many varieties.

* *Felis Catus. F. cauda elongata fufco annulata, corpore fafciis nigricantibus; dorfalibus longitudinalibus tribus; lateralibus fpiralibus.* Lin. fyft. 62. Faun. fuec, N° 9. Br. Zool. i. 69. De Buffon, vi. tab. ii. Briffon quad. 191.

5   Doctor

C A T.

Doctor *Sparman*, p. 148, informs us that he shot a wild cat near the hot baths at the *Cape*, which was of a grey color, and three times the weight of the tame sort. Its length was twenty-one inches: of the tail thirteen. It was exactly the same as the domestic kind; possibly of the same extraction.

α ANGORA CAT. *Schreber*, cvii. B. With long hair; of a silvery whiteness, and silky texture; very long, especially about the neck, where it forms a fine ruff: the hairs on the tail very long, and spreading: is a large variety: found about *Angora*; the same country which produces the fine-haired goat, p. 62. Degenerates after the first generation, in our climate. A variety of this kind is found in *China*, with pendent ears, of which the *Chinese* are very fond, and ornament their necks with silver collars. They are cruel enemies to rats. Perhaps the domestic animals which the *Chinese* call *Sumxi* *.

β TORTOISE-SHELL CAT: black, white, and orange. Le chat d'Espagne. *De Buffon*, vi. *tab.* iii.

γ BLUE CAT. Le chat des chartreux. *De Buffon*, vi. *tab.* iv.

This variety is properly of a dun color, or greyish black. It is much cultivated in *Siberia*, on account of its fine fur; but was brought there, as well as the other domestic kinds, by the *Russians*.

δ The long-headed cat with a sharp nose, from *New Spain*, of

* *De Buffon*, Supplem. iii. 116.

the

Japan Cat ___ . *N.º 196* .

the fize of a common cat: fhort legs: weak claws: round and flat ears, and of a reddifh yellow color; and of a tame nature—is another animal little known*.

The cat, a ufeful, but deceitful domeftic: when pleafed, purrs, and moves its tail: when angry, fpits, hiffes, ftrikes with its foot: in walking, draws in its claws: drinks little: is fond of fifh: the female very falacious; a piteous, jarring, fqualling lover: its urine corrofive: buries its dung: the natural enemy of mice; watches them with great gravity; does not always reject vegetables: wafhes its face with its fore feet, *Linnæus* fays, at the approach of a ftorm; fees by night: its eyes fhine in the dark: its hair emits fire, when rubbed in the dark: always lights on its feet: proverbially tenacious of life: very cleanly; hates wet: is fond of perfumes; *marum, valerian, catmint.* The unaccountable antipathy of multitudes! beloved by the *Mahometans:* *Maillet,* who fays that the cats of *Ægypt* are very beautiful, adds, that the inhabitants build hofpitals for them†.

Chat fauvage Indien. Vofmaer.

196. JAPAN.

C. with upright pointed ears: color of the face and lower part of the neck whitifh: breaft and lower belly a clear grey.: body, part yellow and clear grey, mixed with black difpofed in tranfverfe rays. Along the back, quite to the tail, is a broad band of black: it alfo extends over the upper part of the tail; the lower part femi-annulated with black and grey.

---

* *Seb. Muf.* i. 76. tab. xlvii. fig. i. † *Voy. d'Egypt.* 30.

Size of a common cat : tail ten inches and a half long : is said to be of gentle manners. Its cry is the mewing of a great cat. By Mr. *Vosmaer's* epithet it seems a native of *Japan*.

197. BLOTCHED. Blotched weesel. *Hist. quad.* ed. i. N° 222. Viverra tigrina. *Schreber*, tab. cxv. Chat-Bizaim. *Vosmaer*.

C. with a round head : short nose : pointed ears : white whiskers : yellowish white nose and cheeks ; a round black spot on each side of the former : a dusky line down the middle of the forehead : back and outside of the limbs a reddish brown : sides and thighs yellowish white, blotched with deep brown : tail as long as the body ; of a reddish brown color ; marked spirally near the end with black. Size of a cat.

MANNERS. On re-consideration of this animal, I am induced, not only by its form, but also its manners, to transfer it to this genus. It purrs and murmurs like a cat: its manners are also treacherous; but its appearance in general gentle.

PLACE. It inhabits the neighborhood of the *Cape* of *Good Hope*, and is much sought after for its skin. *Kolben* says it scents of musk, and that it is called the *Biguam* cat. He gives a figure of it, which, like all his others, is very bad. It is of the size of our tame cat.

Felis

Felis Guigna. *Molina Chili.* 275.

C. of a tawny color, marked with round black spots, five lines in diameter, extending along the back to the tail: size of the common cat.

Inhabits *Chili,* and inhabits the forests.

Felis Colorolla, *Molina Chili.* 275.

C. of a white color, marked with irregular spots of black and yellow: the tail encircled with black quite to the point.

This, like the other, inhabits the forests of *Chili:* lives on birds and mice; and sometimes infests the poultry yards. A character of these two species is the having the head and tail larger in proportion than the common cat.

Le chat sauvage de la Nouvelle Espagne. *De Buffon,* Supplem. iii. 227. tab. xliii.

C. with small eyes: tail the shortest, in proportion, of any of this division of the genus: color of a cinereous blue, marked with very short streaks of black: hairs strong enough to make pencils with firm points.

Length four feet; height three.

Inhabits *New Spain.* Described by M. *de Buffon* from a draw-

Q q 2   ing.

ing. He fuppofes it to be the fame with N° 202, the *Serval*; but it is nearly double the fize. The fpots in this are long, in the other round; and if we may credit the drawing, the legs in this are plain, in the *Serval* fpotted.

The *Tepe Maxtlaton* of *Hiſpaniola*, deſcribed by *Seba*, i. 77. *tab.* xlviii. fig. 2. may be referred to this ſpecies.

* With ſhort tails.

Lynxes.

201. MOUNTAIN.

Le Chat-pard. *Memoires pour ſervir a l'hiſt. Nat. An. part.* i. 110.
Catus Pardus ſive Catus Montanus *Americanorum*. The Cat a mountain. *Raii ſyn. quad.* 169.
Felis Pardalis. F. cauda elongata, cor-pore maculis ſuperioribus virgatis, inferioribus orbiculatis. *Lin. ſyſt.* 62. *Briſſon quad.* 199.
Chat ſauvage de la *Caroline. De Buf-ſon*, Supplem. iii. 226. LEV. MUS.

C. with upright pointed ears, marked with two brown tranſ-verſe bars: color of the head, and whole upper part of the body, reddiſh brown, marked with long narrow ſtripes on the back; and with numerous round ſmall ſpots on the legs and ſides: the belly whitiſh: the chin and throat of a pure white: the tail barred with black: the length of this animal two feet and a half; that of the tail eight inches.

Inhabits *North America*: grows very fat: is a mild and gentle animal. The *Quauhpecotli** of *Mexico* agrees in nature with this: is of a brown or duſky color, darkeſt about the back, and gloſſy:

* *Hernandez An. Mex.* 6.   *Seb. Muſ.* i. 68. *tab.* xlii. *fig.* 2.

feet

feet black: on the belly the hair is long and white: difagrees
with the former in the tail, which is thick and long.

Le Serval. *De Buffon*, xiii, 233. *tab.* xxxiv. *Schreber*, cviii.   202. SERVAL.

DIFFERS from the preceding in thefe particulars: the orbits are white: the fpots on the body univerfally round: in
its nature very fierce, and untameable: inhabits the woods in the
mountanous parts of *India:* lives in trees, and fcarcely ever defcends on the ground, for it breeds in them: leaps with great
agility from tree to tree: called by the natives of *Malabar*, the
Maraputé; by the *Portuguefe*, the *Serval**.

203. LYNX.

Chaus. *Plinii lib.* viii. *c.* 19. Lupus
cervarius, *c.* 22
Λυγξ. *Ælian. lib.* xiv. *c.* 6. *Oppian
Cyneg.* iii. 84
Lupus cervarius, Lynx, Chaus. *Gefner
quad.* 677. 678.
Lynx five Leuncia. *Caii opufc.* 50. *Fa-
bri Exp. An. Nov. Hifp.* 5?7.
Lynx, Catus cervarius, *Anglicè*, the
Ounce. *Raii fyn. quad.* 166. *Tourne-
fort's voy. 4to.* i. 300
Rys, Oftrowidz. *Rzaczinfki Polon.* 222.
Srcheber, cix.

Lux. *Kramer Auftr.* 311. *Ridinger
Wilden Thiere*, 22. *Klein Thiere*, 65.
&c.
Felis Lynx. F. cauda abbreviata; apice
atra, auriculis apice barbatis. *Lin.
fyft.* 62.
Warglo, Kattlo. *Faun. fuec.* N° 10, 11.
Lynx Felis auriculorum apicibus pilis
longiffimis præditis, caudâ brevi.
Briffon quad. 200. Catus cervarius,
199.
Le Lynx, ou Loup-Cervier. *De Buffon,*
ix. 231. *tab.* xxi. LEV. Mus.

C• with a fhort tail, black at its end: eyes of a pale yellow:
hair under the chin long and full: hair on the body long
and foft, of a cinereous color, tinged with red, marked with
dufky fpots, more or lefs diftinct in different fubjects; in fome
fcarcely vifible: belly whitifh: ears erect, tufted with long black
hairs, the character of the different fpecies of *Lynxes:* legs and

* *De Buffon.*

feet very thick and ſtrong: the length of the ſkin of a *Ruſſian* lynx, from noſe to tail, was four feet ſix inches; the tail only ſix: vary ſometimes in their color: the *Irbys*, from lake *Balckaſh**, or the *Kattlo* of the *Swedes*, is whitiſh, ſpotted with black, and larger than the common kind; this large variety is called by the *Germans*, *Wolf-Lucks*, and *Kalb-Lucks*, on account of its ſize. In the BRITISH MUSEUM are two moſt beautiful ſpecimens, ſaid to have been brought from *Spain*.

Perhaps it was a variety of this which Doctor *Pallas* informed me was killed in the pine woods, on the banks of the *Volga*, below *Caſan*. It was of an uniform whitiſh yellow above, and unſpotted; beneath white: the ears tipped with black. That might alſo be the variety ſeen by Doctor *Forſter*, in the Empreſs's menagery at *Peterſburgh*, brought from the kingdom of *Tibet*. With duſky ſpots on a yellowiſh white ground; and of a fierce and piercing aſpect.

Inhabits the vaſt foreſts of the north of *Europe*, *Aſia*, and *America*†, not *India*, though poets have harneſſed them to the chariot of *Bacchus*, in his conqueſt of that country: brings two or three young at a time: is long-lived: climbs trees: lies in wait for the deer, which paſs under; falls on them, and ſeizing on the jugular vein, ſoon makes them its prey: will not attack mankind; but is very deſtructive to the reſt of the animal creation. The furs of theſe animals are valuable for their ſoftneſs and warmth: numbers are annually imported from *North America*, and the north of *Europe* and *Aſia*; the farther *North* and *Eaſt* they

---

* Situated weſt of the river *Irtyſh*.

† Wild Cat. *Lawſon Carolina*, 118. *Cateſby App.* xxv. Found as far ſouth as *Mexico*, the *Pinuum Daſypus* of *Nierenberg*, 153.

are

Bay Lynx. — N.º 204.

are taken, the whiter they are, and the more diftinct the fpots;
of thefe the moft elegant kind is called *Irbys*, taken near lake
Balckafh, whofe fkin fells on the fpot for one pound *fterling* *.

The antients celebrated the great-quicknefs of its fight; and
feigned that its urine was converted into a precious ftone †.

> *Victa racemifero Lyncas dedit* INDIA *Baccho :*
> *E quibus (ut memorant) quicquid vefica remifit,*
> *Vertitur in Lapides, et congelat Äëre tacto.*   Ovid. Met. xv. 413.

> *India* when conquer'd, on the conquering god,
> For planted vines, the fharp ey'd *Lynx* beftow'd,
> Whofe urine, fhed before it touches earth,
> Congeals in air, and gives to gems their birth.   DRYDEN.

C. with a fhort tail: irides yellow: ears upright, and fharp-   204. BAY L.
pointed, tufted with long black hairs: color of the head,
back, fides, and exterior parts of the legs, bright bay, obfcurely
marked with dufky fpots: down the face marked with black
ftripes, pointing to the nofe: each fide the upper lip three rows
of minute black fpots, with long ftiff hairs iffuing out of them:
orbits edged with white: from beneath each eye certain long
black ftripes, of an incurvated form, mark the cheeks; which,
with the upper and under lip, whole under fide of the body,
and infides of the legs, are white: the upper part of the infide of
the fore legs marked with two black bars: upper part of the tail
barred with dufky ftrokes; and next the end, one of a deep black;

* *Ritchkoff's Orenb. Topog.* i. 296.   † *Plinii lib.* viii. *c.* 38, xxviii. *c.* 8.

its tip and under fide white.   About twice the bignefs of a large cat: the hair fhorter and fmoother than that of the laft.

Inhabits the inner parts of the province of *New York*.

205. CASPIAN L.	*Chaus* animal feli affine. *Nov. Com. Petrop.* xx. 483. tab. xiv.

L. with a round head, a little more oblong than that of the common cat: fhining reftlefs eyes, with a moft brilliant golden pupil: nofe oblong: the upper lip bifid: whifkers fcarcely two inches long: ears erect, oval, and lined with white hairs; their outfide reddifh; their fummits tufted with black.

Hairs coarfer than thofe of the cat or common *Lynx*, but lefs fo than thofe of the wolf: fhorteft on the head; on the top of the back above two inches long: the color of the head and body a yellowifh brown, or dufky: the breaft and belly of a bright brown, nearly orange: in the infide of the legs, near the bending of the knee, are two tranfverfe obfcure dufky bars: the feet like that of a cat, cloathed with hair, black below.

The tail reaches only to the flexure of the leg, is thick and cylindric, of the fame color with the back, tipped with black, and thrice obfcurely annulated with black near the end.

In general appearance it has the form of the domeftic cat: its length is two feet fix from the nofe to the bafe of the tail: its tail little more than eleven inches: its height before nineteen inches; behind twenty.   It is fometimes found larger, there being inftances of its reaching the length of three feet from the nofe to the tail.

We,

P.Mazell Sculp

Persian Lynx. N.º 207

We are indebted to Mr. *Gueldenstaedt*, who very ably fills one of the professor's chairs in the academy at *Petersburgh*, for the discovery of this animal. It inhabits the reeds and woods in the marshy parts that border on the western sides of the *Caspian* sea, particularly about the castle *Kislar*, on the river *Terek*, and in the *Persian* provinces of *Ghilan* and *Masenderan*, and frequent about the mouth of the *Kur*, the antient *Cyrus*. PLACE.

In manners, voice, and food, it agrees with the wild cat. Conceals itself in the day, and wanders over the flooded tracts in search of prey: feeds on rats, mice, and birds, but seldom climbs trees: is excessively fierce, and never frequents the haunts of mankind: is so impatient of captivity, that one which was taken in a trap, and had one leg broken, refused for many days the food placed by it; but in its rage devoured the fractured limb, with pieces of the stake it was fastened to; and broke all its teeth in the phrenzy of its rage. MANNERS.

Siyah-Ghush, or Black-ear. *Charleton Ex.* xiv.
21. *tab. pag.* 23. *Raii syn. quad.* 168.   Le Caracal. *De Buffon,* ix. 262. *tab.* xxiv.   207. PERSIAN.
Ph. Transf. vol. li. part ii. 648. *tab.*   *Schreber,* cx. LEV. MUS.

C. with a lengthened face, and small head: very long, slender, black ears, terminated with a long tuft of black hairs: inside and bottom of the ears white: nose white: eyes small: the upper part of the body is of a very pale reddish brown: the tail rather darker: belly and breast whitish: limbs strong, and pretty long: the hind part of each marked with black: tail about half the length of the body.

Inhabit *Persia, India,* and *Barbary**: are often brought up

* *Shaw's travels,* 247. The mouth of the *Barbary* variety is black, and the face fuller.

     R r   tame,

tame, and used in the chace of lesser quadrupeds, and the larger sort of birds, such as cranes, pelicans, peacocks, &c. which they surprise with great address: when they seize their prey, hold it fast with their mouth, and lie for a time motionless on it: are said to attend the lion, and to feed on the remains of the prey which that animal leaves *: are fierce when provoked: Dr. *Charleton* says, he saw one fall on a hound, which it killed and tore to pieces in a moment, notwithstanding the dog defended itself to the utmost.

The *Arabian* writers call it *Anak el Ard:* say that it hunts like the panther; jumps up at cranes as they fly; and covers its steps when hunting †.

6. LIBYAN.

C. with short black tufts to the ears, which are white within; of a lively red without: tail white at the tip, annulated with four black rings, with the same black marks behind the four legs.

Greatly inferior in size to the former; not larger than a common cat. Inhabits both *Libya* and *Barbary* ‡.

* *Voy. de Thevenot,* iii. 204. The *Arabs,* according to *Thevenot,* call it *Kara-Coulac,* or Black-ear.

† Dr. *Thomas Hyde,* in *Ulugh Beigh, tab. p.* 36. The figure is from an original drawing by Mr. *Edwards.*

‡ *De Buffon,* Supplem. iii. 232. from Mr. *Bruce.*

END OF THE FIRST VOLUME.

Printed in the United States
By Bookmasters

Betriebssysteme